Solution Protocols to Festering Island Disputes

Since the coming into force of the United Nations Law of the Sea, states have been targeting outlying islands to expand their exclusive economic zones, simultaneously stirring up strident nationalism when such plans clash with those of neighbouring states. No such actions have brought the world closer to the brink of war than the ongoing face-off between China and Japan over the Diaoyu/Senkaku islands, an uninhabited archipelago in the East China Sea.

In this timely and original book, Godfrey Baldacchino provides a detailed exploration of seven tried and tested solution protocols that have led to innovative 'win–win' solutions to island disputes over the last four centuries. A closer look at the circumstances and processes that brought contending regional powers to an honourable, even mutually advantageous, settlement over islands provides a convincing and original argument as to why the conflict over the Diaoyu/Senkaku islands need not conclude in a 'zero-sum' or 'winner takes all' solution, as is the likely outcome of both open conflict and international arbitration.

The book will be of interest to scholars and practitioners concerned with the festering Diaoyu/Senkaku dispute, as well as students, scholars and policy specialists in geography, geopolitics, international relations, conflict studies, island studies, Asian studies and history.

Godfrey Baldacchino is Pro-Rector (International Development) and Professor of Sociology at the University of Malta and UNESCO Co-Chair (Island Studies and Sustainability) at the University of Prince Edward Island, Canada. He is also President of the International Small Islands Studies Association, Chair of the Scientific Board of RETI (The Network of Island Universities) and Founding Executive Editor of *Island Studies Journal*.

Solution Protocols to Festering Island Disputes

'Win–Win' Solutions for the Diaoyu/Senkaku Islands

Godfrey Baldacchino

With contributions from
Christian Fleury, Adam Grydehøj, Pamila Gupta, Steven Hillebrink, Akihiro Iwashita, Henry Johnson, Mari Katayanagi, Bjarne Lindström, Liu Jiangyong, David McIntyre, Elisabeth Nauclér, Peter van Aert and Jouke van Dijk

Routledge
Taylor & Francis Group
LONDON AND NEW YORK

First published 2017 by Routledge

2 Park Square, Milton Park, Abingdon, Oxfordshire OX14 4RN
52 Vanderbilt Avenue, New York, NY 10017

Routledge is an imprint of the Taylor & Francis Group, an informa business

First issued in paperback 2018

British Library Cataloguing in Publication Data
A catalogue record for this book is available from the British Library

Library of Congress Cataloguing in Publication Data
A catalog record for this book has been requested

ISBN: 978-1-4724-7518-3 (hbk)
ISBN: 978-0-367-21877-5 (pbk)

Typeset in Times New Roman
by Keystroke, Neville Lodge, Tettenhall, Wolverhampton

Contents

Maps

Contributors

Godfrey Baldacchino is Pro-Rector (International Development) and Professor of Sociology at the University of Malta and UNESCO Co-Chair (Island Studies and Sustainability) at the University of Prince Edward Island, Canada. He is also President of the International Small Islands Studies Association, Chair of the Scientific Board of RETI (The Network of Island Universities) and Founding Executive Editor of *Island Studies Journal*.

Christian Fleury has a PhD in geography and is research fellow at ESO Caen (University of Caen-Normandie, France). His research interests concern frontier studies, marine spaces and geopolitical disputes, coastal societies and 'island laboratories'.

Adam Grydehøj holds a PhD in ethnology, is the Director of Island Dynamics, lead editor of the *Urban Island Studies* journal, and a Research Associate at the University of Prince Edward Island, Canada.

Pamila Gupta is Associate Professor at WISER (Wits Institute for Social and Economic Research) at the University of Witwatersrand in Johannesburg, South Africa. She is interested in the contribution of islands to the construction of modernity.

Steven Hillebrink is a Legal Officer at the Directorate for Constitutional Affairs and Legislation of the Dutch Ministry of the Interior and Kingdom Relations in The Hague, the Netherlands.

Akihiro Iwashita is Professor, History and Area Studies/Slavic Studies, Hokkaido University, Japan. His research focuses on border studies in Eurasia, including Russia, China and Central Asia.

Henry Johnson is Professor, Department of Music, University of Otago, New Zealand. His areas of expertise include Asian studies and the anthropology of Japan. He is originally from Jersey, Channel Islands.

Mari Katayanagi is Professor at the Graduate School for International Development and Cooperation, Hiroshima University, Japan. She has served as Senior Advisor at the Embassy of Japan in Bosnia and Herzegovina, a political advisor at the Office of the High Representative (Bosnia and Herzegovina) and more recently as Research Fellow at the Japan International Cooperation Agency (JICA) Research Institute.

Bjarne Lindström is an economic geographer and former director of the Agency of Statistics and Research in the Åland Islands (ÅSUB). He is currently working on a senior sub-consultancy for Sweco Society Ltd in Stockholm, Sweden.

Liu Jiangyong is Vice-Director and Professor of International Relations at the Institute of Modern International Relations, Tsinghua University, Beijing, People's Republic of China. His research focuses on Japanese political and foreign policy and international relations in the Asia-Pacific region.

David McIntyre served as Professor of History at the University of Canterbury, New Zealand. He has a special interest in Pacific island societies and politics.

Elisabeth Nauclér is a Swedish-born jurist and Director General of the Åland Government (1999–2006). She was the single Member of Parliament representing the autonomous Åland Islands in the Parliament of Finland (2007–2015).

Peter van Aert is a Faculty Member at the Institute for Culture, Society & State, National University of Tierra del Fuego, Argentina. His research focuses on migration, cultural configuration and local development.

Jouke van Dijk is the Chair of the Board of the Waddenacademie, Leeuwarden, the Netherlands, as well as a Professor of Regional Labour Market Analysis and Chair of the Department of Economic Geography, Faculty of Spatial Sciences, University of Groningen, the Netherlands. He is also President of the European Regional Science Association (ERSA).

Preface

I must have been still an undergraduate student at the University of Malta, in Malta, in the late 1970s when I attended a guest lecture by Ewan Anderson, then Professor of Geopolitics at the University of Durham, United Kingdom, on the theme of political hot spots around the world: the theme of his (then forthcoming) book, *Atlas of world flashpoints* (1993). I was immediately and totally hooked by the excitement and dangers of such volatile corners of the world. I bravely walked up and exchanged some observations with Prof. Anderson after his illuminating presentation. I still have his business card. I wondered for some time why he, or anyone really, would want to go through the hassle of two PhDs.

The idea of so many places at, or almost at, war intrigued me. And as I shifted my attention to the study of islands late in my academic career, I realised that many small islands, inhabited or otherwise, were indeed bones of contention, disputed and claimed by multiple powers. And, in spite of their obvious small size, and often unimpressive mineral or other natural resources, countries were willing to dedicate installations, military equipment, even lives, to stake and defend their claims to such seemingly innocuous places.

There is one other compelling reason why I decided to take on this book. I am sometimes challenged about the relevance and rigour of island studies in dealing with hard topics. It may be fine, it is said, to deal with island literature or island tourism or island environments, but how helpful is the 'island lens' in understanding island politics, economic development or international relations? My answer has always been an incontrovertible 'yes'; and I have set about proving this to be so through my own scholarship. In a similar vein, I could not give up the opportunity when, in 2012, I saw that China and Japan were on the brink of war because of a 'bunch of rocks', and very few of the analysts poring over the affair were bothering to reflect on how other 'small island' stories could throw some fresh and badly needed light and inspiration on the tense goings on in the East China Sea.

Here, then, is hopefully more proof of the bearing and power of an island studies epistemology to what appear to be intractable border disputes.

Godfrey Baldacchino
5 August 2016

Acknowledgements

In previous books, I have left her for last; but that is not quite fair. So, this time, I start by unabashedly thanking Anna, my wife of 30-plus years, for patiently supporting my academic adventures and accompanying me on my visits to China (and Taiwan) and Japan (and Okinawa). None of these academic quests, nor the ensuing scribbling and writing, would have been possible, or proved as fruitful, without her patient and constant support. (It is my turn to happily support her now, as she progresses on her own PhD journey.)

I thank Professor Kawai Kei, Director, Professor Shunsuke Nagashima and the rest of the staff of the Research Centre for the Pacific Islands at Kagoshima University, Kyushu, Japan – 国際島嶼教育研究センター – for hosting me as Visiting Research Professor for seven weeks (January to March 2015) while I worked on the research that came to shape the foundation for this book. I also thank Professor Huei-Min Tsai, Taiwan National Normal University, Taipei, Taiwan; Professor Mari Katayanagi, The Graduate School for International Development and Cooperation (IDEC), Hiroshima University, Hiroshima, Japan; Professor Yoko Fujita, Director of the International Institute for Okinawan Studies, University of the Ryukyus, Okinawa, Japan; Professor David McMurray, Head of the International Studies Department, International University of Kagoshima (IUK), Japan; and many other associates and friends, including Professor Lawrence M. Liao, Hiroshima University, Japan, who facilitated my presentations and research gathering in China and Japan during spring 2015. I am grateful to Mr Reno Calleja, President of the Malta–China Friendship Society and former Minister of Tourism in Malta (and a former student of mine), for introducing me to H. E. The Ambassador of the People's Republic of China to the Republic of Malta, Mr Cai Jinbiao, and H. E. The Ambassador of the Republic of Malta to the People's Republic of China, Mr Clifford Borg-Marks. Through their kind interventions, my wife Anna and I were the guests of the Chinese People's Institute of Foreign Affairs (CPIFA), Beijing, China – 中国人民外交学会 – for a week in March 2015, where we had an intensive programme of meetings and lectures with senior officials, policy makers and academics; my special thanks also to Ambassador Sun Rongmin, Vice-President of the CPIFA, Dr Lu Jintao, Director (European Section) of the CPIFA, and a very patient Ms Ni Chenxi who coordinated our visit flawlessly. I thank Stephen A. Royle, outgoing professor of Island Geography at Queen's

University Belfast, Northern Ireland, UK, for critical comments on an earlier draft of my first manuscript to explore the Diaoyu/Senkaku affair, and to Dr Andrew McGregor (past editor) and reviewers of *Asia Pacific Viewpoint* who eventually piloted that manuscript to publication in that reputable journal (and on whose international advisory board I serve).

I must also thank the various expert contributors who have enriched this book with their detailed knowledge of particular island disputes: Christian Fleury, Adam Grydehøj, Pamila Gupta, Steven Hillebrink, Akihiro Iwashita, Henry Johnson, Bjarne Lindström, David McIntyre, Elizabeth Nauclér, Peter van Aert and Jouke van Dijk are the legitimate co-authors of this publication and I am truly grateful for their contributions. Akihiro Iwashita (again), Mari Katayanagi and Liu Jiangyong kindly provided the introductory and concluding thoughts to this volume. I also thank Lorne Kriwoken at the University of Tasmania, Australia, for his expert opinions on the Antarctica section of Chapter 9; and Adam Grydehøj (again) for sharing many of the ideas that are found in Chapter 13.

At Routledge, my publisher, I extend my gratitude to Faye Leerink, Commissioning Editor, and Priscilla Corbett, Editorial Assistant, as my main contacts. They have wholeheartedly believed in this book project and provided every encouragement and support, even while Ashgate – with whom the book project was kicked off – was being acquired by Routledge/Taylor & Francis. I also thank my employer, the University of Malta, in Malta, for granting me the paid research leave during 2015 that permitted the scoping and drafting of this manuscript. I also thank the University of Prince Edward Island, Canada, for engaging me as a sessional lecturer during the summers of 2015 and 2016: it was then, and in the salubrious location of Charlottetown, PEI, my second home, that most of the writing of this manuscript has taken place.

The three articulated maps, showing the location of the Diaoyu/Senkaku islands at various scales, has been kindly drawn by Nikoleta Koukourouvli, at the Laboratory of Local and Island Development, University of the Aegean, Lesvos, Greece.

I remain responsible for any errors of omission or commission, and for which I apologise in advance. The usual disclaimers apply.

Moreover, I wish to gratefully acknowledge various publishers and copyright owners for permission to reproduce the following material that appears in this text:

From Chapter 5

Material adapted from 'Autonomy without sovereignty: an island studies approach': Chapter 1 of the text: G. Baldacchino (2010). *Island enclaves: Offshoring strategies, creative governance, and subnational island jurisdictions*. Kingston, ON and Montreal, QC, Canada: McGill-Queen's University Press. © McGill-Queen's Universtiy Press.

From Chapter 7:

Hillebrink, S. (2013). 'Saint-Martin/Sint Maarten'. In G. Baldacchino (ed.), *The political economy of divided islands: Unified geographies, multiple polities* (pp. 176–194). New York: Palgrave Macmillan. © Palgrave Macmillan.

Iwashita, A. (2013). Boris Ussurriiski/Teixihazi. In G. Baldacchino (ed.), *The political economy of divided islands: Unified geographies, multiple polities* (pp. 212–226). New York: Palgrave Macmillan. © Palgrave Macmillan.

From Chapter 8:
McIntyre, W. D. (2012). The partition of the Gilbert and Ellice Islands. *Island Studies Journal*, *7*(1), 135-146. © Institute of Island Studies, University of Prince Edward Island, Canada.

From Chapter 10:
van Aert, P. (2016). The Beagle conflict. *Island Studies Journal*, *11*(1), 307–314. © Institute of Island Studies, University of Prince Edward Island, Canada.
Fleury, C. and Johnson, H. (2015). The Minquiers and Écréhous in spatial context: contemporary issues and cross perspectives on border islands, reefs and rocks. *Island Studies Journal*, *10*(2), 163–180. © Institute of Island Studies, University of Prince Edward Island, Canada.

From Chapter 12:
Gupta, P. (2015). Futures, fakes and discourses of the gigantic and miniature in "The World" islands, Dubai. *Island Studies Journal*, *10*(2), 181–196. © Institute of Island Studies, University of Prince Edward Island, Canada.

From Epilogue I:
Liu Jiangyong (2013). The Diaoyu/Senkaku island dispute and future development of Sino-Japanese relations. In Institute for Strategic Studies, National Defence University of the People's Liberation Army (ed.), *International strategic situation and China's national security*. People's Republic of China: Military Science Publishing House. © The author.

From Epilogue II:
Iwashita, A. (2015). *Japan's border issues: Pitfalls and prospects*. New York: Routledge. © Routledge/Taylor & Francis.

The arguments that populate the pages of this book have also been made, though much more succinctly and timorously, in: Baldacchino, G. (2016). Diaoyu Dao, Diaoyutai or Senkaku? Creative solutions to a festering dispute in the East China Sea from an 'island studies' perspective. *Asia Pacific Viewpoint*, *57*(1), 16–26. With thanks to *APV* Editor Lisa Law and Wiley for permission to reproduce. © 2016 John Wiley and Sons.

A note on style

References to Chinese names respect the convention in that country, whereby family names appear before first names. In the case of such Chinese names, these are provided in full in the list of references, as applicable.

Wherever possible, I refer to the Diaoyu/Senkaku islands in this way, with the Chinese version of their name first and the Japanese second, in strict alphabetical order. (The Taiwanese Diaoyutai – or Tiaoyutai – is a variant of the Chinese name; while Chinese references to Diaoyu Dao are, strictly speaking, only to the main island of the group, known as Uotsori-shima to the Japanese.) Some of my co-contributors and I have also referred to the Diaoyu islands singly, or to the Senkaku islands singly, when making the case on behalf of the official Chinese and Japanese government positions respectively. Others may have preferred Senkaku/Diaoyu or Diaoyu/Senkaku to signal their support for, or the respective strength of, particular claims.

I use the same, strict alphabetical sequence to refer to another contested island chain: the Falklands/Malvinas.

No reference, argument or any other content in this volume is to be interpreted as suggesting that any island, or group of islands, belongs to one claimant state. It is not the purpose or competence of this book to adjudicate or assign any island to any country.

Abbreviations

AD	Anno Domini (Christian Era)
ADIZ	Air-Defence Identification Zone
AIIB	Asian Infrastructure Investment Bank
AOSIS	Alliance of Small Island States
APEC	Asia-Pacific Economic Cooperation
ASEAN	Association of South-East Asian Nations
ATS	Antarctic Treaty System
BBC	British Broadcasting Corporation
BC	before Christ (before Christian Era)
CBM	confidence-building mechanism
CIA	Central Intelligence Agency (US)
COP	Conference of the Parties
CPC	Communist Party of China
CPSL	Trilateral Working Group on Coastal Protection and Sea Level Rise
CRAMRA	Convention on the Regulation of Antarctic Mineral Resource Activities
CWSS	Common Wadden Sea Secretariat
DMZ	Demilitarised Zone, Korea
DPJ	Democratic Party of Japan
DPP	Democratic People's Party of ROC/Taiwan
ECSPI	East China Sea Peace Initiative
EEC	European Economic Community
EEZ	Exclusive Economic Zone
EPZ	Export Processing Zone
EU	European Union
GDP	gross domestic product
GEIC	Gilbert and Ellice Islands Colony
IAATO	International Association of Antarctica Tour Operators
IANA	Internet Assigned Numbers Authority
ICJ	International Court of Justice
ICZM	Integrated Coastal Zone Management
IGY	International Geophysical Year
IPCC	Intergovernmental Panel on Climate Change

IUCN	International Union for the Conservation of Nature
JPDA	Joint Petroleum Development Area (Timor Sea)
KMT	Kuomintang (中國國民黨: Nationalist Party of ROC/Taiwan)
LDP	Liberal Democratic Party, Japan
LMS	London Missionary Society
LOSC	Law of the Sea Convention
MPA	Marine Protected Area
MSA	Maritime Safety Agency, Japan
MSDF	Maritime Self-Defense Force, Japan
NGO	non-governmental organisation
NJ	State of New Jersey, USA
NY	State of New York, USA
OST	Outer Space Treaty
PEP	Protocol on Environmental Protection (ATS)
PLA	People's Liberation Army, PRC
PLN	Picton, Lennox and Nueva Islands
PRC	People's Republic of China
RIB	rigid hulled inflatable boat
ROC	Republic of China (Taiwan)
SAR	Special Administrative Region
SFG	L'Espérance International Airport (known as Grand Case), French Saint-Martin
SIDS	small island developing states
SMX	Princess Juliana International Airport, Dutch Sint Maarten
SNIJ	sub-national island jurisdiction
SNJ	sub-national jurisdiction
TBPA	transboundary protected area
TSDA	Timor Sea Designated Authority
TST	Timor Sea Treaty
TWSC	Trilateral Wadden Sea Cooperation
UAE	United Arab Emirates
UK	United Kingdom
UN	United Nations
UNCLOS	United Nations Conference on the Law of the Sea
UNEP	United Nations Environment Programme
UNESCO	United Nations Educational, Scientific and Cultural Organization
USA/US	United States of America
USSR	Union of Soviet Socialist Republics
WHS	World Heritage Site (UNESCO)
WSF	Wadden Sea Forum
WWI	World War One
WWII	World War Two

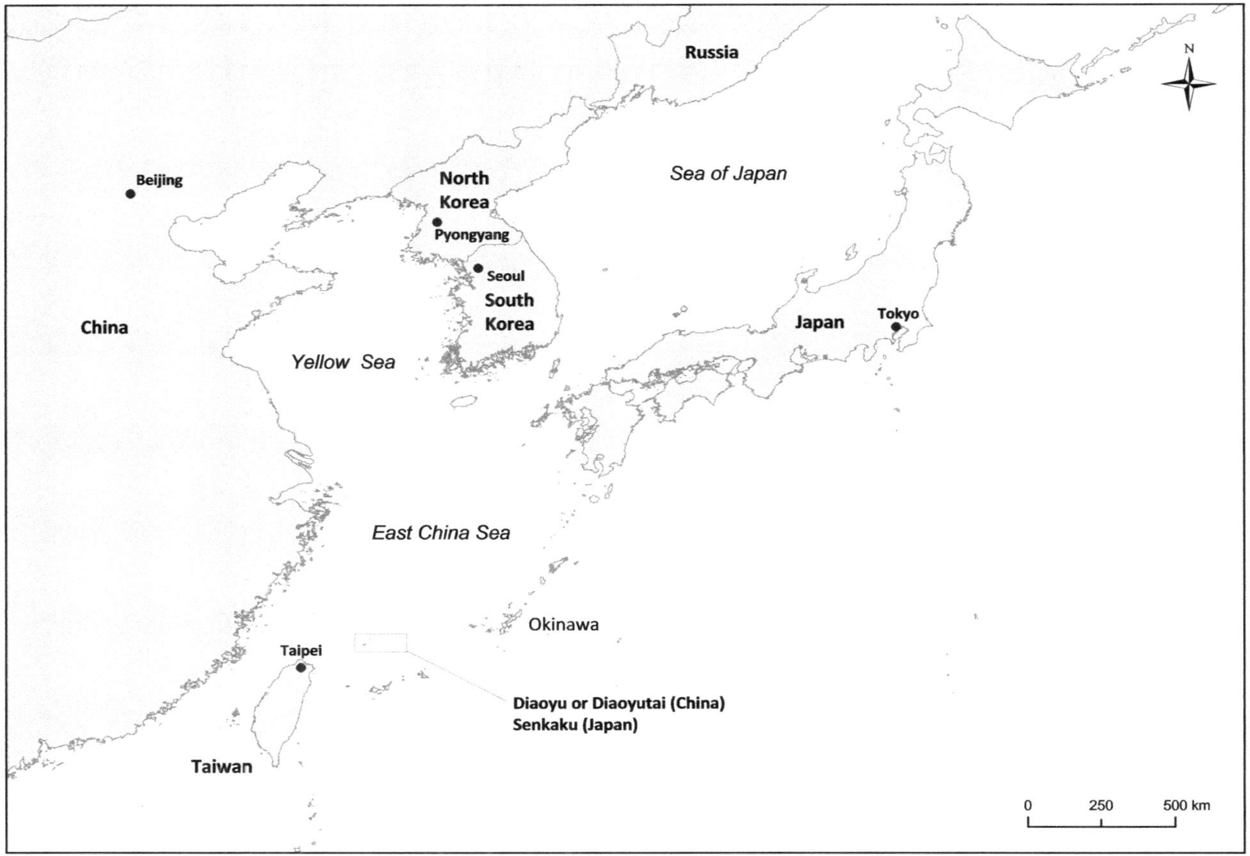

Map 1 East Asia, with a focus on the East China Sea

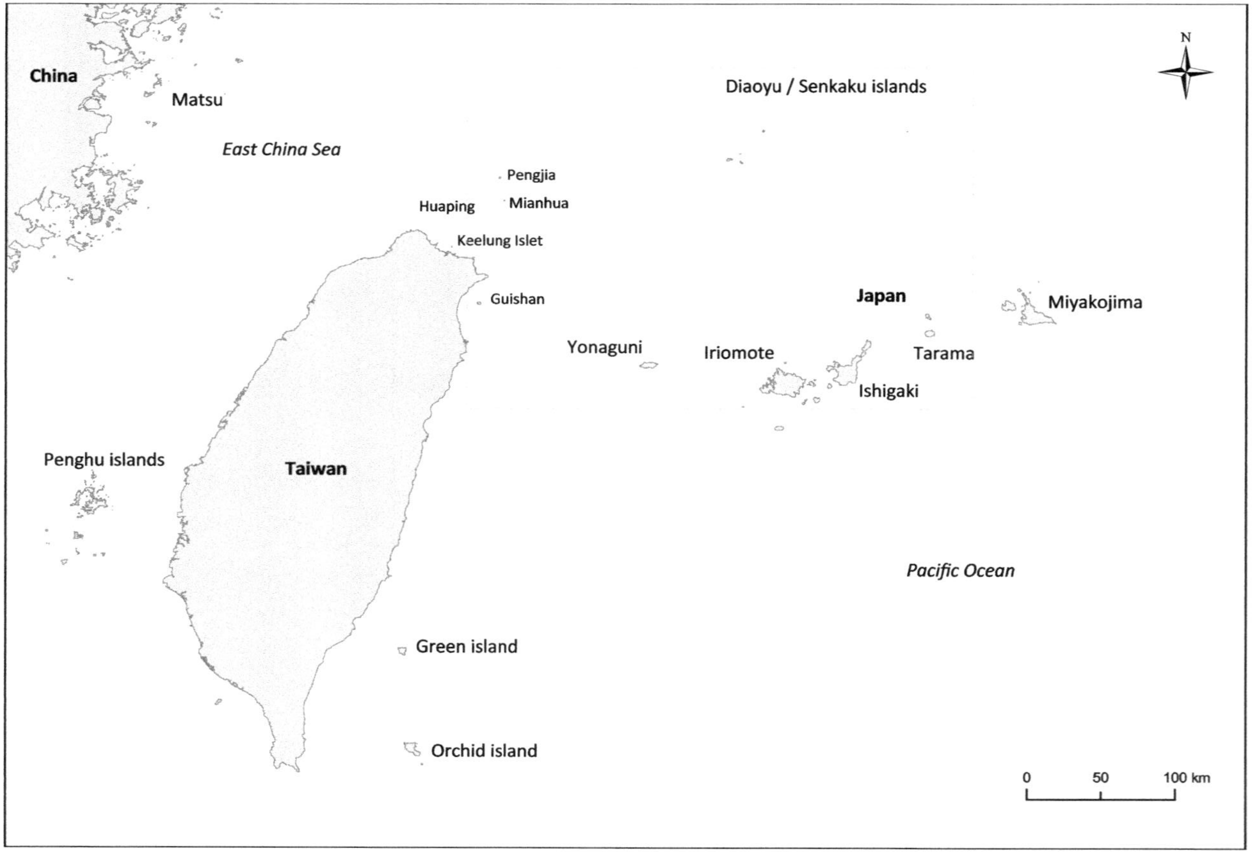

Map 2 The Diaoyu/Senkaku island group, along with neighbouring islands of Okinawa Prefecture, Japan, and of China (including Taiwan)

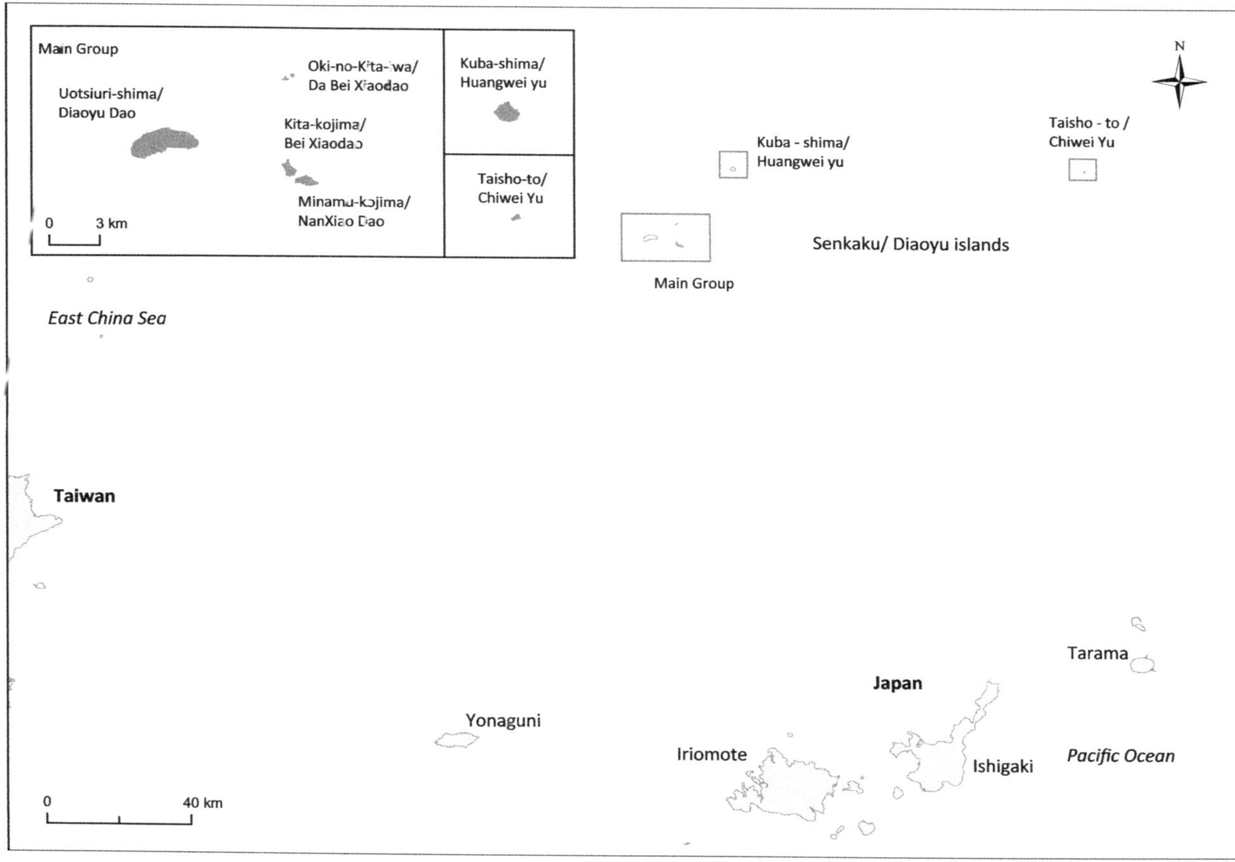

Map 3 The Diaoyu/Senkaku archipelago, showing its named constituent members

Prologue

Mari Katayanagi

Introduction: that elusive truth

Is truth single and unique? In countries that have experienced internal strife, including an internal armed conflict, each of the groups aligned on different sides to a hot argument often believes in a different 'truth'. One of the reasons for this phenomenon is that each group may have received different information before, during and after the conflict. Not knowing what the other side actually has done or does, or hearing only a part of their acts, a negative image of 'the other' can easily be created, and becomes very hard to dislodge. In media studies, it is said that we have a tendency to listen only to what we want to hear. Unwanted information thus passes by, unfiltered and unnoticed or demeaned.

Truth commissions – or their variations, such as truth and reconciliation commissions and truth and justice commissions – are devices designed to get closer to the truth of what may have actually happened during a violent conflict, particularly covering systematic human rights violations. The actual experience, however, tells us how difficult it is to reconstruct a truth that can be perceived as such by all the sides who may have fought against each other. Once a negative image of 'the other' is formed during a conflict, it can be hardly changed. Likewise, strong beliefs in the notion of a just war, meant to protect national pride or 'defend the motherland', tenaciously remain in contemporary society.

When Professor Baldacchino offered me this opportunity to make this modest contribution of mine to this fascinating book, I was not aware of the level of the proximity between the issues surrounding the Diaoyu/Senkaku islands and what my research usually focuses on, which is peacebuilding in conflict-affected countries and regions. Recent armed conflicts are more likely to take place inside the boundaries of a country, whereas the Diaoyu/Senkaku issue is a burning one between two neighbouring, sovereign states, Japan and China. Fortunately, it has also not (yet) triggered an armed conflict. It is interesting to see the issue in comparison with the peacebuilding or conflict resolution measures of a divided society.

Peacebuilding and conflict resolution

In the case of the Diaoyu/Senkaku issue, the two governments and their academics have been arguing historical truth, presenting 'evidence' from each viewpoint respectively. As a citizen of Japan, I am almost certain that most of the people in both countries do not know the details of these arguments. The vast majority would probably pick up bits and pieces mainly from media reports, and – at best – form a kind of view or opinion on the matter. It is thus crucial what the media reports, and whether we have an open mind to listen to what we are not interested in hearing. In other words, both sides should listen to what the other side says without any prejudice. Nonetheless, the arguments of experts on both sides are quite technical, and therefore I am very much sceptical that non-experts are able to form any justifiable conclusions or informed opinions based on the limited knowledge from media reports.

As can be read in greater detail in this timely book, Japan has three major unresolved territorial disputes: the Northern Territories issue with Russia, the Takeshima issue with the Republic of Korea and the Senkaku/Diaoyu issue with China. However, the Japanese government recognises the existence of disputes only with regards to the first two. For the Japanese government, the Senkaku/Diaoyu affair is not an issue, as the following quote from the website of the Ministry of Foreign Affairs of Japan explains:

> There is no doubt that the Senkaku Islands are clearly an inherent part of the territory of Japan, in light of historical facts and based upon international law. Indeed, the Senkaku Islands are under the valid control of Japan. There exists no issue of territorial sovereignty to be resolved concerning the Senkaku Islands.
>
> (Ministry of Foreign Affairs, Japan, 2016a)

It is with some irony that such a categorical and dogmatic statement should appear on a page titled 'Japanese Territory', where two other disputes are also explained. Obviously, there would not be any need for such a strong declaration if nobody questions the islands' sovereignty. When it comes to the Takeshima issue, the situation is reversed: the Republic of Korea denies the existence of any dispute; but Japan insists that one exists. Only the Northern Territories are recognised as disputed by both sides, in this case Russia and Japan.

Conflict resolution requires both parties to acknowledge the existence of one or more problems and discuss them. Without dialogue, the problems are left to their own devices and some of them can trigger violent confrontation. In peace-building activities, international peacebuilders often organise meetings of former adversaries. Some representatives may initially reject meeting former enemies and peacebuilders may seek to persuade them to attend a meeting. The ensuing personal communication may eventually transform and mellow the relationship.

The nationalisation of the Senkaku islands by the Japanese government in 2012 led to an unexpected turn of events, not only for China but for Japan itself as well. The decision was mobilised by the then governor of Tokyo, who was notorious for some quite disgraceful statements, and who had announced a plan for the purchase of the Senkaku islands by the Tokyo Metropolitan Government. The then Japanese Prime Minister Noda, however, considered that the matter was not for a local government to decide arbitrarily but for the central government. The act of nationalisation proper, however, outraged the Chinese government, as detailed in this book, since it was an implicit act of sovereignty by Japan and thus a blatant affront to the Chinese claims over the same islands. Relations between the two countries significantly cooled. The rise in tension not only affected the diplomatic and official sphere of Sino-Japanese relations, but also the private sector and academia. (On this point, some statistical data will be discussed below.) The dominance of a political event over the daily life of citizens is a threatening phenomenon. The political relationship deprives ordinary citizens of opportunities to develop and strengthen trust between the two countries. Much social, economic and cultural 'bridging capital' built painstakingly over the years can be, and was, suddenly lost because of a political crisis that resonated with deep-rooted historical suspicions and animosities.

Is the issue of the uninhabited Diaoyu/Senkaku archipelago, 'the five islets and three rocks', relevant to the daily life of people? Possibly for some fishers in the region, and potentially for more, if natural resources are to be eventually extracted from surrounding waters. However, it is hard to identify any immediate relevance with wider publics today. Does it then make any sense to heighten the tension between the two neighbouring countries and even consider an armed confrontation, possibly heralding a Third World War, as some have warned?

Towards a peaceful settlement?

The maintenance of security and peace has been the primary concern of international relations. The peaceful resolution of conflict is a foundational principle enshrined in the United Nations Charter. What, then, can we foresee as a way towards the non-violent settlement of the Diaoyu/Senkaku dispute?

Generally speaking, one way is to seek a regional solution beyond bilateral negotiation. A Japanese scholar, Kimie Hara, proposes the utilisation of a multilateral framework which leads to mutual concessions and collective gains. She suggested two hypothetical concession frameworks, trilateral on the one hand, and a four-party one on the other. The former can be a combination of Japan's concession to Korea regarding Dokdo/Takeshima, China's concession to Japan as to Senkaku and Okinawa, and Korea's concessions concerning the naming issues of surrounding seas. Russia comes into the scene in the latter framework. The idea is to add a two-island transfer settlement to the above, meaning the mutual concessions between Russia and Japan. In her view, what is apparently an intractable 'win–lose' situation in a bilateral context can be perceived somewhat differently and in a more positive sense in a multilateral framework (Hara, 2012). Each party gains something and offers something else. While it may not sound immediately feasible in this case, a regional solution may be a possibility in other cases.

Another approach is to seek a solution under international law. Taking the particular case of the Diaoyu/Senkaku islands before an international tribunal has also been discussed and proposed by a number of scholars. However, we have recently witnessed the outcome of such an option in another case. The July 2016 ruling of the Permanent Court of Arbitration in the 'Philippines versus PRC' case about the South China Sea is a good example to reflect on the approach. International arbitration is certainly a means of conflict resolution under international law. Yet, even before the judgment, China's reaction to the ruling was to reject its lawfulness, declaring it null and void. For the Chinese government, however, this does not mean that it will neglect international law. The government further declared that it 'will continue to abide by international law and basic norms governing international relations as enshrined in the Charter of the United Nations, including the principles respecting state sovereignty and territorial integrity and peaceful settlement of disputes' (*Shanghai Daily*, 2016). Manila is showing a readiness to promote dialogue with Beijing, while keeping the 2016 ruling as the basis for such a dialogue. The ruling may have positive effects for peaceful dispute resolution in the region; not because its legal force is recognised by China, but thanks to such a considerate approach by the Philippines government, combined with the diplomatic pressure of the international community towards China to respect international law. Nevertheless, this example makes us wonder whether an international tribunal would be an effective option to address the Diaoyu/Senkaku issue. We should remind ourselves that the Japanese government does not recognise the existence of such a dispute.

One more note regarding legal solutions: we are inclined to acknowledge authority in such terms as 'legal', 'illegal', 'lawful', 'unlawful' and so on, quite often without deeply analysing the discussion and meaning behind these terms. Using such language simply for the purpose of undermining the other side's argument may work against the resolution of the problem, unless the analysis behind the argument is capable of being accepted by the other side.

The Diaoyu/Senkaku issue comes with a political and emotional baggage. Since 2012, it is clearly deemed as one of the most serious problems pertaining to the history of relations between the two nations, a situation that a number of scholars had feared could arise. The competition to present historical proof of sovereignty and related public relations activities can hardly lead to a resolution of the complex issue. We will need a conflict resolution or transformation approach and, for that, a formal 'shelving' option – a mutual agreement not to touch the sovereignty question for a certain period – may be the way forward. By doing so, the two countries can engage in constructive discussions, such as joint resource management or the development of a bi-territorial nature reserve.

Meanwhile, a creative solution over sovereignty can continue to be studied, and many rich examples of such possible 'solution protocols' are presented in this book: co-habitation on a divided island, a condominium with a periodic alternating governance, a territory for commercial activity open to treaty signatories, a shared jurisdiction, a divided jurisdiction and so on. The crucial lesson that we should learn from history is to respect human lives, and this requires the prevention of armed conflicts. This book imparts various means and creative thinking to us, underlining the significance of a 'win–win' solution.

The figures speak for themselves

To build and foster friendly relations between countries, ties between their respective publics would help to erase prejudice and dilute much of strident nationalistic propaganda and rhetoric that abounds on the Internet. In this sense, it is encouraging to note that, even during the heat of antagonism over the Diaoyu/Senkaku affair, many Chinese students were studying in Japan; indeed, the largest number of non-Japanese students in Japan by nationality is Chinese (JASSO, 2016). This number did not decrease, but increased, after the collision of the Chinese trawler with Japanese patrol boats in 2010; while it decreased in 2013 by approximately 4,400, possibly due to the tension after the act of nationalisation of the Senkaku islands by Japan. However, the number quickly recovered in 2014 and remained stable at just over 94,000 Chinese students studying in Japan in 2015. When it comes to business visitors, although the figure appears relatively flat, a significant increase (more than 48,000 persons) in 2010 turned to a decrease in 2011, but the figures rebounded in 2012, and an overall 65 per cent increase – from some 182,000 to 270,000 business visitors annually – has been registered in the period 2009–2014 (JNTO, 2016). Turning next to the number of tourists, the negative consequences appear manifest at first sight. The number of Chinese tourists visiting Japan recorded a staggering 46 per cent decrease from 2010 to 2011 – but the main reason for this was probably the devastating 11 March 2011 Tohoku earthquake and tsunami (WTTC, 2011) – and a 15 per cent decrease from 2012 to 2013. However, the overall figures demonstrate clearly that strong ties have been developing steadily between the two countries in multiple spheres, although political turmoil may cause temporary setbacks: the number of Chinese tourists to Japan rose four-fold from 482,000 in 2009 to 1,754,000 in 2014 (JNTO, 2016).

In the search for mutual benefits of both countries, people from various walks of life can contribute to a peaceful resolution of the Diaoyu/Senkaku affair. It is known that the private sector has taken important initiatives for conflict resolution recently in two other islands: Ireland and Sri Lanka. The more people engage in forming personal ties with the other side, the more possibility we will have to transform the relationship between the two nations, so that a creative solution for non-violent settlement comes true.

Conclusion

With this book, Professor Baldacchino offers us an interestingly refreshing perspective to the Diaoyu/Senkaku affair. He reminds us that this is, and remains, essentially a festering dispute over a group of small and uninhabited islands; and so, it behoves us to examine similar island cases from the past for possible inspiration. Many such cases may have appeared intractable and non-negotiable at certain historic moments, and of course they would have lent themselves to different versions of 'the truth', supported readily by different shreds of 'evidence'.

And yet, lasting solutions have been found for many of these island disputes. Indeed, we may be lucky that such islands, large and small, have been privileged candidates to various creative solution-making throughout history, dating back at least to the division of the island of St Martin in 1648.

I would probably be speaking on behalf of many by looking forward to the day when the Diaoyu/Senkaku islands join such a list of successful, 'win–win' resolved disputes.

1 Introduction

Glorious mission update

> As Chinese, we should unite together to defend the Diaoyu Islands and completely defend our homeland. Let us annihilate any attempt by the Japanese devils to come ashore our islands, throw them back to the Japanese isles, defend the Diaoyutai.
>
> (Introduction to *Defend the Diaoyu Islands*, game on iPad)

At least, in a corner of the hyper-world of video gaming, the matter of the Diaoyu Islands – also known as Senkaku to the Japanese and as the Diaoyutai in Taiwan – has been resolved. A video game for China's iPad users launched in 2012 featured much of the vitriol that has been flung back and forth between China and Japan over a series of small islands and rocks in the East China Sea during that same year. In this game, called *Defend the Diaoyu Islands*, players are challenged to attack and resolutely evict Japanese 'invaders' by force from a group of small islands that are considered to be an unalienable part of China, doing so with a variety of weapons and military material, including a deployment of the talismanic *Liaoning*, the first aircraft carrier commissioned for the People's Liberation Army (PLA).

The game is an update of *Glorious Mission Online*, a gaming platform originally part-sponsored by the Chinese PLA as a virtual simulator to train soldiers, and fashioned after the popular *Call of Duty* game series developed in the United States. The game is hopping on the '*kang ri*' or 'anti-Japanese' bandwagon: a term that refers to a recent raft of Chinese dramas, movies and games that invariably feature Japanese characters as the villains of the piece.

A video trailer for the new downloadable content was posted to YouTube (Blum, 2013a). An English caption reads, 'Diayou Island is inherent territory of the Chinese nation', while hawkish images of Japanese Prime Minister Shinzo Abe flash across the screen. A montage of fighter jets, parachuting soldiers, tanks and naval gunships with cannons firing quickly follows. A picture of the disputed island chain then appears. 'Response to defend the Motherland' is displayed on the next screen as the background music intensifies. Suddenly, various computer graphic angle shots of the *Liaoning* carrier appear. Toward the end of the trailer, machine gun-wielding PLA soldiers burst into a Japanese encampment, with bullets flying. A small rising sun banner pops up on-screen for each virtual

Japanese soldier that is neutralised. In classic *Call of Duty* fashion, an approaching attack helicopter is blown away by a bazooka shot (Blum, 2013b). The stereotyped Japanese 'enemies' include moustachioed World War Two soldiers, samurai, clumsy sumo wrestlers and ninja warriors. There are altogether 71 levels in the game, each named after an island.

The game developers are Shenzhen Zhongqingbaowang Interaction Network Co Ltd, or ZQGame. This is a Chinese company that makes massively multiplayer online games as well as other browser and mobile games (ZQGame, 2016). *Defend the Diaoyu Islands* is not their first foray into plugging into the Chinese citizens' apparent zeal for conflict with their neighbours. The company's website includes several promotions for online chapters of the company's banner game set in China during World War Two, including one called *Nanjing Trial* that puts gamers in the city of Nanjing during the infamous Nanjing Massacre to fight off the intruding Japanese. The game's tagline spells out the content clearly: 'Return to Nanjing, pay back blood with blood.' In *Resistance War Online*, another game created by Shenzhen ZQGame in 2007, the advertising slogan reads: 'Kill the devils if you are Chinese!' The 'devils', of course, are Japanese: anti-Japanese war dramas make up a third of the series shown on Chinese TV. In present-day China, the quantity of anti-Japanese war movies and teleplays is abnormally high, and their values deeply enmeshed in a radical nationalism, resulting is a general trend towards a 'carnival of vengeful images' (Gai, 2015). One recent offering is *Royalty in Blood*, a grisly 36-part TV series about the Sino-Japanese conflict of 1937–1945, now downloadable off the Internet (Love TV Show, 2015).

The emergence of patriotic digital games in China can been interpreted as both reactions to the perceived rising wave of government-led militarism in Japan, as well as attempts to profit from the nationalistic impulse of an increasingly gaming-savvy Chinese public. The Chinese government has cleverly integrated online game technology into an expanding and modernising propaganda tool, utilised it for disseminating official ideology, sustaining economic growth in the high value e-gaming sector, while also offering venues for the display of (electronic, and therefore rather politically safe) nationalism. Through policy regulations, financial support and partnership with private game companies, the Chinese Party-state has turned online gaming into a profitable industry, a vehicle for the projection of soft power and an instrument for patriotic education (Nie, 2013).

Such 'projections' sometimes go too far: China has had to rein in excessive anti-Japanese sentiment on various occasions. In August 2004, riots broke out after a Japanese victory over China in the Asian Cup Soccer Final in Beijing. Vandalism started outside the stadium after the game; and security forces were at hand to contain the damage (CNN, 2004). When Tokyo pursued membership of the UN Security Council early in 2005, an online anti-Japan petition against the proposal was signed by millions of Chinese; streets protests called for a boycott of Japanese products, and vandalism broke out in various Chinese cities, including Shanghai (Marquand, 2005). The Diaoyu/Senkaku affair would catalyse even larger protests in 2012 (about which read more in Chapter 4).

Anti-Japanese protests are nothing new in China. Such and similar episodes of (often grassroots) anti-foreign mobilisation in China have been tolerated or repressed by the PRC to signal either resolve or reassurance (Chen Weiss, 2014). In fact, *Defend the Diaoyu Islands* was taken down from the China Apple store without warning (Waugh, 2012). One explanation for the sudden removal is that the game violated Apple's terms of service: its rules specify that 'enemies' within the context of a game cannot target a specific and actual race, culture, government or corporation, or any other real entity (Martin, 2012).

In any case, the game was good while it lasted: in the first week after its launching, *Defend the Diaoyu Islands* stormed to the top ten free iPad games available off the China iTunes app store. Most game downloaders come from China, the United States and, perhaps surprisingly, Japan. Or, perhaps not: a game is a game after all.

Passion clouding pragmatism beyond the fantasy world of video games

The same burning questions can be raised, beyond the fantasy world of video games: to what extent have the statements and manoeuvres indulged in by China (and Taiwan) and Japan around a group of small, uninhabited (and probably uninhabitable) islands to be taken seriously? To what extent should one believe the official rhetoric that these barren rocks are inviolate sovereign territory, as claimed by each of these two neighbouring sovereign states? And to what extreme lengths would each of the two countries be prepared to go to defend such a claim? Moreover, which of the two countries has the strongest claim to the islands?

Much has been written in response to the escalation witnessed in the East China Sea in recent years involving the Diaoyu/Senkaku Islands. For a few weeks in September 2012, the world held its breath: few could believe that China and Japan, both among the world's three largest economies and with a total bilateral trade of almost US$334 billion in 2012, could actually go to war – and over a heap of small rocks. To get a sense of how small, check out the complete list of islets and rocks that make up the archipelago known as Diaoyu or Senkaku, with their respective land area (see Table 1.1).

Table 1.1 The eight main members of the Diaoyu/Senkaku island chain, with land area

Japanese name	*Chinese name*	*Land area (km^2)*
Uotsuri-shima (魚釣島)	Diaoyu Dao (釣魚島)	4.32
Kita-kojima (北小島)	Bei Xiaodao (北小島)	0.3267
Minami-kojima (南小島)	Nan Xiaodao （南小島）	0.4592
Taishō-tō (大正島)	Chiwei Yu (赤尾嶼)	0.0609
Kuba-shima (久場島)	Huangwei Yu (黄尾嶼)	1.08
Oki-no-Kita-iwa (沖ノ北岩)	Da Bei Xiaodao (大北小島/北岩)	0.0183
Oki-no-Minami-iwa (沖ノ南岩)	Da Nan Xiaodao (大南小島/南岩)	0.0048
Tobise (飛瀬)	Fei Jiao Yan (飛礁岩/飛岩)	0.0008

Source: Kakazu (2016, pp. 2–3)

For Japan, struggling to emerge from two decades of economic malaise, exports to burgeoning China are a key source of growth. Companies from Sony to Toyota desperately need Chinese consumers to buy their television sets and automobiles – which, by the way, may be both assembled by Chinese workers in China – to offset a sluggish home market. Meanwhile, China imports more from Japan than from any other country, and many of those 'made-in-Japan' goods – such as high-tech components and capital equipment – are critical to China's economic advance. And yet, passions over the Diaoyu/Senkaku islands have been clouding pragmatism. In China, sales of Japanese-branded cars plummeted amid large anti-Japan protests in 2012. Japanese direct investment into China plunged by nearly 37 per cent in the first nine months of 2013 (Schuman, 2013). By 2014, the value of bilateral trade had still not returned to pre-2012 levels. 'The fledgling trade recovery between Asia's two economic giants could be derailed by another political dispute' (Obe, 2014). In its cover for the 22–28 September 2012 edition, the seriously popular *The Economist* magazine – which claims to reach 5.3 million readers a week in print and online – offered readers a view of the disputed islands, suitably scaled down to highlight their small size, with the question printed in bold type: 'Could China and Japan really go to war over these?' A green turtle, swimming in the bottom right corner of the cover, ventures to answer, 'Sadly, yes'.

The widespread perception of the very real dangers of a confrontation that could even usher in a World War Three has catalysed a very large corpus of literature on the topic. Chinese, Japanese and third-party experts, researchers and scholars have published many titles on the case, and continue to do so. These provide details of the historical antecedents to the affair; and propose arguments and counter-arguments as to who of the two countries might have the stronger claim at international law. Above all, most seek to locate the whole matter within a strategic and complex macro-political analysis: the change of leadership in the Chinese Communist Party and the consolidation of the position of the General Secretary of the Communist Party of China, the President of the People's Republic of China and the Chairman of China's Central Military Commission, Xi Jinping; Japanese Prime Minister Shinzo Abe's legacy and Japan's growing militarisation; what is perceived as the shifting power balance in East Asia, given the rising and more assertive China; the recalibration of US power in the region as part of its 'pivot' to Asia; and a lingering, heavy, historical memory that ensures that there is no love lost between the two countries (Christensen, 2011).

This book

This text takes a different tack. It acknowledges that the matter at hand is essentially a disagreement over and about *islands*, and there is therefore scope and promise in reminding ourselves that, as such, the Diaoyu/Senkaku affair shares a pedigree with other island rows that have flared up in the course of history. These have typically involved two, often neighbouring powers, keen to secure territorial and/or marine rights (such as to fishing and/or seabed or underwater mineral resources), to project national power by the extension of exclusive economic zones (EEZs), and to stoke local national sentiment and fervour at home. Quite typically, the confrontation did not deal with the disputed islands per se. More, much more, was at stake. So far, the comparisons with the Diaoyu/Senkaku stack up nicely.

But, we can go further: history affords us the advantage of hindsight. From the comfortable perch of the twenty-first century, we can look back – at least 400 years – and coolly assess how, and why, some (though not all) of these squabbles over islands have been 'solved', in some way or other. A raft of solutions has been found to assuage the parties involved in these island clashes, enough to establish a *modus vivendi* that does not risk escalating into all-out violence and warfare. For those concerned with the Diaoyu/Senkaku affair, it would be opportune, wise and timely to note these many cases of resolution, since any, or a combination, or a hybrid amalgam of their details, could well provide a viable roadmap for what could unfold in the East China Sea, and indeed even for other, presumably more intractable, situations involving islands elsewhere – such as the South China Sea (Cronin et al., 2012).

A regime for island disputes: UNCLOS

This text is also original in another sense. After all, many interpreters of the Diaoyu/Senkaku quarrel do invoke other islands in order to shed light on the case; but they do so in the context of the track record of the outcome of international arbitration. In other words, these commentators source legal decisions that have outcomes which have (largely) determined who has title over what. (We will consider some of the most interesting exceptions in the chapters that follow.) The implicit understanding of such island–island comparisons is that such episodes provide glimpses into zero-sum game scenarios – who might be *right* and who might be *wrong* in claiming title to the Diaoyu/Senkaku islands – and not in *how such title may and can be creatively shared.*

The principal mechanism for the adjudication of island disputes similar to the Diaoyu/Senkaku affair would now be the International Court of Justice (ICJ). And, apart from the ICJ's own case law, the legal coda on which such adjudications would be based derive from the United Nations Conference on the Law of the Sea (UNCLOS), a treaty that both China and Japan signed in 1996 (UNCLOS, 1982). It was the ambassador of a small island state to the UN that kick-started the processes that eventually led to UNCLOS. In a speech to the UN General Assembly in 1967, Malta's Ambassador to the United Nations, Arvid Pardo, asked for 'an effective international regime over the seabed and the ocean floor beyond a clearly defined national jurisdiction' (UNCLOS, 1998, n.p.). What started as an exercise to regulate the seabed turned into a global diplomatic effort to regulate and write rules for all ocean areas, all uses of the seas and all of its resources, both in the sea and below the seabed. These factors eventually led to the convening of the Third United Nations Conference on the Law of the Sea (UNCLOS III), to write a comprehensive treaty for the oceans.

UNCLOS was adopted by 119 nations in Montego Bay, Jamaica, on 10 December 1982. The Convention came into force on 16 November 1994, 30 days after the 60th nation had ratified it. By August 2016, the time of writing, a total of 164 countries (plus Palestine, Cook Islands and Niue, which have observer status at the UN) and the European Union had ratified the Convention. The Convention contains 320 articles that regulate coastal and island nations' rights and duties connected to their coastal sea areas.

Exclusive economic zones (EEZs) are determined by criteria as set forth by UNCLOS III. These zones may be based on a 200-nautical mile (370.4 km; or 230.2 miles) extension from the coast, or may extend further if based on a longer continental shelf. Japan claims the border of its EEZ to lie at an equidistant point between the coasts of its southern prefecture of Okinawa and that of mainland China; while Beijing claims that its EEZ extends to the edge of its continental shelf, intruding well into the median line as claimed by Tokyo. In such claims and measurements under UNCLOS, states may stake claims to the extent of territorial seas, contiguous zones and EEZs with baselines departing from their island territories. Article 121(3) specifically states that a rock that is unable to either sustain 'human habitation' or 'economic life' may *not* be used to create an EEZ or continental shelf: there are sufficient grounds to believe that both China and Japan think that possession of the Diaoyu/Senkaku Islands will widen their respective EEZ claims to some degree.

What UNCLOS *does* allow is for the joint development of marine resources that lie in contested territory. (This is one 'solution protocol' that will be explored further in this text.) Otherwise, the public understanding is that the Diaoyu/Senkaku islands *must* belong to either one side or the other, even though there is a track record of international agreements (mostly beyond UNCLOS) that has creatively offered alternative scenarios of dispute resolution. Other than testing the relevant case law to see who is right or wrong (about which more below), many international observers are keen to advise the obvious 'confidence-building mechanisms' (CBMs), direct hotlines for top level communication and bilateral agreements for managing military behaviour: presumably, there is no disagreement and much common sense on advocating such measures (Smith, 2014b). But, while such measures may calm tense situations, the actual causes of the dispute are not addressed and do not go away.

Case law of island disputes: seven rulings

When dealing with the legal niceties of the Diaoyu/Senkaku affair, scholars and experts look around for similar and comparative case law. Seven such island rulings are usually invoked, the latter five adjudicated by the ICJ, the principal judicial organ of the United Nations (UN). These are: the island of *Palmas* (a dispute between the Netherlands (for what is now Indonesia) and the US (for what is now the Philippines)) awarded in 1928 (UN, 2006); *Clipperton* island (between France and Mexico) in 1931 (ILSA, 2012); the *Minquiers* and *Écréhous* (between the United Kingdom and France) in 1953 (ICJ, 1953), about which more in Chapter 10; the islands of *El Tigre*, *Meanguera* and *Meanguerita* as part of a broader settlement between Honduras, El Salvador and Nicaragua, in what is known as the Gulf of

Fonseca Case, in 1992, about which more in Chapter 8 (ICJ, 1992); the island known as *Kasilili* in Botswana and Sedudu in Namibia, in 1999 (ICJ, 1999a), about which more also in Chapter 10; the islands of *Pulau Ligitan* and *Pulau Sipadan* in a dispute between Indonesia and Malaysia in 2002 (ICJ, 2002); and the islands of *Pedra Branca, Pulau Batu Puteh, Middle Rocks* and *South Ledge* in the case between Malaysia and Singapore, in 2008 (ICJ, 2008). These seven cases are nicely complementary; and they provide a consistent and self-referencing body of law governing island disputes. It is their cumulative case law that would provide an obvious framework in which to analyse claims to other disputed islands or island chains, such as (but not exclusively) the Diaoyu/Senkaku islands.

What this corpus of island dispute case law suggests is that, generally, the claim of having discovered *terra nullius* alone does not grant actual title to the discoverer. (*Terra nullius*, Latin for 'nobody's land', is used in international law to describe territory which has been relinquished by, or has never been subject to the sovereignty of, any state.) Discovery only conveys inchoate title, which requires more concrete acts of actual occupation or an exercise of authority to demonstrate sovereignty over an island. A state that has inchoate title over territory may lose sovereignty over such territory to another state by *prescription*, if the latter state actually occupies or exercises state authority over the said contested territory. Moreover, claims to uninhabited islands require exercising less authority over the island to maintain title than an inhabited island; however, a state must still protest the acts of other competing states that constitute occupation or the exercise of state functions in timely fashion, or that claimant state risks losing title altogether. Additionally, historical or documentary evidence will not trump the actual long-term, peaceful occupation or the exercise of state authority over a territory in dispute. Moreover, besides actual occupation, the ICJ or an arbitrator will also consider whether an interested party essentially acquiesced to the other state's occupation of a territory: the evidence of effective possession and control of an island by one party without protest or action by the other constitutes what is known in legal jargon as *estoppel*, and points to acquiescence (Heflin, 2000; Miyoshi, 2014, p. 16). Lastly, exercising exclusive control over visits to an island; conducting detailed surveys; installing and maintaining beacons, lighthouses, and/or radio communications equipment for use by one's naval forces; and soliciting official bids from private citizens and entrepreneurs to conduct commercial activity on the same island, *as long as* such activity is undertaken on the basis of official regulations or under governmental authority, are all undertakings that help to establish prescriptive title (Scoville, 2014, p. 583; 2015, pp. 99–100).

The overall thrust of these seven decisions suggests to many that, should China and Japan ever agree to take the Diaoyu/Senkaku matter to arbitration, the evidence appears to be stacked in Japan's favour; and this factor introduces another strong reason as to why China might not want to consider arbitration (Dai, 2006, pp. 149–154; Miyoshi, 2014). Nevertheless, others argue that Japan's claim to sovereignty of the Diaoyu/Senkakus may be less firmly grounded in international law than meets the eye (e.g. Lee and Ming, 2012).

And yet, even were the parties to seek an adjudication of the dispute in some judicial form (such as the ICJ) or in a quasi-judicial context (as in an arbitration), it remains uncertain if the decision makers would secure compliance with the outcome, should the latter assign the Diaoyu/Senkaku *en bloc* to one party in its eventual ruling. The conflicting states have much at stake, including quantifiable mineral and economic resources, as well as large doses of nationalist historiography and priceless national pride. Moreover, their governments would likely be prone to considerable popular pressure to disregard the diktat of a foreign body/third party: this has indeed already occurred in East Asia, in a similarly contentious territorial dispute, involving Cambodia and Thailand (*International Herald Tribune*, 2008). Indeed, it seems most likely that *no* ruling will carry much if any legitimacy if this grants full and unencumbered sovereignty, and the associated access to resources, to just one claimant party (Harris, 2013, p. 609). Particularly on the part of China, perceptions of the fairness and legitimacy of (Western) international law and justice can be quite fickle (Franck and Franck, 1995), and especially if they do not tally with the espoused nationalist narrative.

Beyond zero-sum games

History also teaches us another lesson: there really need not be such a thing as a zero-sum game. It is often our limited imagination, the shackles of the past, our unwillingness to be creative and a stunted entrepreneurship that stop us from considering those options, taking those steps and reaching those decisions that secure 'win–win' solutions. And here, small islands loom as problematic spaces. Their inherent compact physicality and their absolutely small size seem to suggest that there can only be one victor in the contest. 'The winner takes it all / the loser standing small', as the Swedish group ABBA's song lyrics warn (Anderson and Ulvaeus, 1980). And, more dangerously, media outlets have taken up thus dubious mantra and proclaimed it as dogma. For example:

> When it comes to maritime disputes, islands, not oil, are the greater threat to international stability. Why? Because resources can be shared *but islands cannot*. In a winner-takes-all environment, leaders have little choice but to dig in their heels. If one country obtains sovereign control over contested territory, the other loses it.
>
> (Meierding, 2015, n.p., my emphasis)

and

> But on some issues, *it is impossible for both sides to win*. China's territorial disputes, for example, with the possible exception of the one with India, which is big enough to allow room for a conceivable compromise, *are zero-sum*.
>
> (Banyan, 2014, n.p., my emphasis)

Such assertions are false. One must be wary and critical of affirmations – such as the ones above – that dismiss creative solutions; or that implicitly expect creative solutions to reside only in large tracts of contested land space. As this book will (hopefully) amply demonstrate, the sheer minor scale and size of small (and especially uninhabited) islands and archipelagos, and their often lack of exploitable, land-based natural resources, may create those very conditions conducive to playful experimentation between contending parties, leading to creative solutions, and which challenge conventional conceptualisations of sovereignty.

I am more than aware that islands, whether large or small, do *not* inspire division. Of the world's thousands of inhabited islands today, *only ten* are split between more than one state: I should know, since I edited a book about this intriguing topic not so long ago (Baldacchino, 2013). And yet, these 'unified geographies but fragmented polities' (ibid.) cannot simply be discarded or dismissed as an odd ragtag set of candidates that justifies some perverse, golden rule of indissolubility. Sharing islands by splitting them physically into plural polities, as these ten islands manifest, may be a rare but still a viable strategy. I am not arguing that splitting does not bring along its own problems; of course it does: consider the intercommunal talks still under way in Cyprus (Ker-Lindsay, 2014); or the revival of the idea of a united Ireland in the wake of the UK referendum on 23 June 2016 to leave the European Union (Humphries and Ferguson, 2016); or the tension surrounding the status of Haitian immigrants in the Dominican Republic (Welsh, 2015). But the borders have held so far, and the arrangements persist, and will probably continue to do so, even out of sheer political inertia, as long as and until a better solution protocol presents itself.

I propose and deploy the term *solution protocols* to broadly describe a cluster of working arrangements, often secured following protracted negotiations and political transactions, and often accompanied by military manoeuvres or threats thereof, and which meet most or all of the aspirations of contending parties. While each is undoubtedly specific to the case and context at hand, the arrangements as a whole establish what come across as innovative yet stable frameworks of tenable and peaceful co-existence, and constitute positive sum, 'win–win' endgames. Such protocols include sharing or splitting sovereignty over islands, suppressing or shelving sovereignty claims, maintaining single sovereignty but sharing jurisdiction, and swapping or selling islands. Over a dozen island examples of these protocols have been compiled and are described in some detail in this book (Chapters 7–12), these in turn being introduced by an examination of how creative governance practices have unfolded, mainly on and for islands (Chapter 6). Their discussion has been entrusted also to experts from specific disciplines and jurisdictions and who are privy to specialised and local knowledge on each of these interesting cases; these esteemed academic colleagues of mine thus enrich this text with their local expertise. The islands showcased in this way range from huge, continental Antarctica (14,000,000 km^2) to tiny Pheasant Island, the world's smallest condominium (6,820 m^2, nearly two acres). To these six solution protocols we add the status quo as the seventh: clearly, the benefits of stalemate or standstill must be acknowledged and explained to fully understand the situation, and realistically appraise the chances of, and risks associated with change (Chapter 13). Finally, Chapter 14 takes us back to the realm of creative governance, but this time looking ahead at future scenarios and their prospects. From this largely speculative exercise, it appears fair to argue that continentalist China is better equipped to live with power asymmetry than exclusively archipelagic Japan is.

Before launching into the core discussion and illustration of solution protocols, the book provides a concise context to the Diaoyu/Senkaku dispute, framing it as part of the current churn in the Asia Pacific: the rise – but actually more of a return – of an assertive China, the enduring question of Taiwan, the recent attempts to bring Japan back to significance, the tensions in Okinawa and the transformation of the US role in the region in the face of all this turbulence (Chapters 2–4). The text also reviews why countries may go, and have gone, to war over small islands, with the Falklands/Malvinas war of 1982 and its aftermath offering a recent and poignant reminder of this situation as well as how win/lose 'solutions' that leave one party smarting from loss are not necessarily the best form of outcome (Chapter 5).

Conclusion: solution protocols and creative governance

The innovative feature of these solution protocols is driven not just by the willingness of the parties to seek a resolution – an outcome *not* to be assumed in the Diaoyu/Senkaku case, however, as shall be argued below – but also by the nature of the island involved *qua* island. The playful experimentation with evolution to which Nature has subjected many (especially oceanic) islands appears to have been copied by modern-day governments who deploy island spaces in *exceptional* ways: not just in the sense of such islands being remarkable spaces, but also in the ways in which states deliberately craft and condone a suspension of the extant law in these small (hence more manageable) locales (Agamben, 2005; Baldacchino, 2010). Think offshore finance centres, nature reserves, theme parks, military bases, detention centres, aboriginal enclaves, free trade zones, gated communities and UNESCO World Heritage Sites *in toto* (Baldacchino, 2006). The list of the special uses to which to put island spaces has grown longer and is more ambitious than the quarantine stations and high-security prison services that small islands have performed for continental powers over many centuries. If the archipelagos of the Galápagos and Hawai'i impress us for their biological diversity and species endemism (Loope et al., 1988), other islands should similarly wow us by their penchant for enclaving, creative governance and jurisdictional novelty. The reductive logic of nation-state container-spaces may have captivated the imagination of (first European and now global) citizenries and governments over time (Johnson, 2012), but it may now be opportune to take stock of the various forms of 'post-sovereign governance' (Karkkainen, 2004) that have been mushrooming suggestively around us, particularly in the conservation of maritime spaces. If the fluidity of trans-boundary waters represents contemporary challenges to, and opportunities for, modern governance (Blatter and Ingram, 2001), then small islands may offer the best hope for bringing these ideas onshore.

Could the Diaoyu/Senkaku islands do the same? If they could join the clutch of islands that have benefitted from creative solutionism, as showcased in this book, then what is currently a bone of contention in the diplomatic, social and economic relationship between China and Japan, and a focus of tension and instability in East Asia, would not just be defused but would also be transformed into an unparalleled opportunity.

A pipe dream? Perhaps. But perhaps not. This book's proposition is not simply an adoption of that insightful adage by German historian Max Weber: believe in and aim for the impossible so that the possible becomes true. It is also one where there is a dire need to showcase and acknowledge those solution protocols, tried and tested in the course of time, which already exist. Let's face it: while the UNCLOS regime has been useful in handling many disputes, it can never exercise a monopoly over options in dispute resolution. Moreover, the UNCLOS provision of distinguishing 'rocks' from 'islands' has been a bone of contention ever since the treaty's inception, since UNCLOS has (so far) failed to define the sustenance of 'economic life' and 'human habitation' that distinguishes one from the other, and whether these need to be viable, self-sustaining and/or independent of outside support: on their premise, huge swathes of ocean territory can be claimed (Haas, 2011). Additionally, when it comes to arbitration, rulings have a habit of riling both sides, since parties do not like returning home with less than anticipated results or outcomes, and certainly not empty-handed. With classical, indivisible concepts of sovereignty, regardless of whether a dispute is resolved peacefully or through armed intervention, one side invariably loses on a claim that it, its leadership and/or its publics, had believed to be valid. The losing side will lick its wounds and work its grudge; waiting for the right moment to pounce back, and possibly doing so, with dramatic results (e.g. Gilbert, 2004). The converse argument also applies: considering and applying re/solution protocols that rely upon non-indivisible notions of sovereignty may be the ideal model for creating durable results that do not erupt into further explosive conflict and dissent down the line (Samuels, 2007, p. 729).

In their own time, these island cases may have appeared to be just as intractable and amenable only to 'win–lose' situations: this is just as the Diaoyu/Senkaku matter appears today. But this need not remain the case. Indeed, looking at the considerable (and still growing) literature about the Diaoyu/Senkaku affair, it appears that, looking at the issue per se, and delving into the historical and political antecedents to the affair, tends to solidify the recognition of the high stakes stand-off. This approach makes it harder, and not easier, to contemplate alternative approaches or initiatives. If, on the other hand, some hard-nosed, evidence-based decision making is required, and if the fragile nature of politics opens a window of opportunity for creative options to policy makers, then this text, premised on a comparative island framework, should offer some fresh and welcome guidelines.

2 Context

Churn in the Asia-Pacific

There is a huge chasm between the possible and the practical. As with the Falklands/Malvinas affair – more about which in Chapter 5 – the Diaoyu/Senkaku case has had various suggested 'solutions'; and more will be provided in this book. At least, tensions between China and Japan have begun to ease after hitting a peak in 2012. Tokyo can now be seen to have acknowledged a dispute over sovereignty of the islands, even if obliquely. A 'four-point agreement', including a commitment to set up crisis management mechanisms, is very welcome (*The Economist*, 2014a). Why are the contending parties so locked in their positions, even when they know that they can each ill afford to antagonise the other? What are the extenuating circumstances that prevent Beijing, Taipei and Tokyo from putting this matter to bed, and may indeed stubbornly lock the matter in crisis mode? A comprehensive answer must first comprise a critical look at the broad political economy of the Western Pacific: regional politics, nation-building, shifting prima donnas and that elephant in the room – US militarism – with various social movements confronting military power in the region in the name of the millions who call the Asia-Pacific their home (Davis, 2015). This review will be followed by more detailed considerations of China and Japan proper in a separate, dedicated chapter.

This sober assessment follows in parallel to the outpouring of many reflections on the legacy of the 'war to end all wars', World War One (WWI), which started in August 1914, and whose 100th anniversary was being remembered at just about the same time that the Diaoyu/Senkaku island affair was making the news headlines. Parallels could not be avoided: 'this kind of fantasising is now all the rage' (Evans, 2014a, n.p.). And yet, such analysis may very well provide the basis for powerful perceptions. And, while perceptions may be false, if people define situations, even false ones, as real, then they are real in their consequences. As Thomas and Thomas (1928, p. 572) remind us: it is perceptions, and not necessarily facts, that typically provide the foundations for belief, decision and action.

The generic optics from the West, and from Japan, regarding China in the twenty-first century, are fairly clear.

In 1904, Sir Halford Mackinder, the founder of geopolitics, announced that the Heartland of Eurasia, the 'World Island' or 'the pivot area', would hold the key to world domination, resolutely eclipsing the power of maritime regimes (Mackinder,

1904). According to Mackinder's analytic lens, Germany/Prussia and the Soviet Union/Russia have already attempted to secure a role of pre-eminence in this Heartland within the last 100 years, but – Russia's intervention in Georgia in 2008 and its annexation of the Crimea in 2014 notwithstanding – their expansionism has been either checked or curtailed by 'the West', though it took two world wars, plus a 'Cold War' (1945–1989) to accomplish this. Is China now the new rising star in the Eurasian heartland? And will its rise and growing assertiveness also lead to such a showdown with Western powers? Does a growing cockiness from China imply a 'rebalancing' of other regional players, including the US? Should China's rise be contained and/or channelled, and, if so, how? The world order today may be far more stable and cooperative than it was in 1914 (Evans, 2014b); a thick web of economic inter-dependence links us all; and all the great powers have nuclear weapons or are allied to someone who does, thus reducing the incentive to strike (Christensen, 2015, p. 290). A cocktail of financial globalisation, unchecked nationalism and escalation-prone conflict may, however, be creating volatile and explosive conditions similar to those that prevailed a century ago.

The great renaissance of China

Already, the story of an armed conflict between the US and China makes a fine topic to a futuristic fictional war novel (Singer and Cole, 2015). If there is 'an enhanced temptation to roll the dice, even though the game may be fatal' (James, 2014), will China be seen to be the one rolling the dice?

In his address as outgoing General Secretary of the Communist Party of China in 2012, Hu Jintao spelt out what a rising China should do, particularly about its maritime policy: 'We should enhance our capacity for exploiting marine resources, develop the marine economy, protect the marine ecological environment, resolutely safeguard China's maritime rights and interests, and build China into a maritime power' (Hu, 2012, n.p.). Developing itself as a maritime power is somewhat of a new endeavour for China. Not since the early fifteenth century has China adopted a deliberate maritime vision: during 1405–1433, the Ming rulers financed seven naval expeditions, designed to establish a Chinese presence and impose imperial control over Indian Ocean trade, impress foreign peoples in the Indian Ocean basin with China's wealth and technological prowess, and extend the Ming empire's tributary system to a hitherto unforeseen expanse (Menzies, 2003). Much has changed since: if China was the pre-eminent kingdom in the region at that time, the country now has to contend with a suite of adjoining and relatively young sovereign states and/or former colonies of European powers (Brunei, Indonesia, Malaysia, the Philippines, Singapore, Vietnam) that have their own maritime aspirations and ambitions. Moreover, Japan and Korea, two neighbouring old regimes, each with a history of tribute to China, are also sovereign and uneasy with China's real intentions about its maritime calling. Look at a map of East Asia, and it is easy to sense that these countries together, literally, stand doggedly in the path of China's putative eastward and southward maritime expansion. China's gritty pursuit of the Diaoyu/Senkaku island affair and, more significantly, its claim to the waters of the South China Sea – islands, reefs and all, within a 'nine-dash line', plus a tenth dash placed near Taiwan – make more sense when seen within this grand scheme. Add to this mix doses of strident populist nationalism, the growth in the number of media outlets through which Chinese citizens can express such views and the increasing sensitivity of the PRC government – indeed, *all* governments – to public opinion. No wonder that Beijing elites feel obliged to react stridently to all perceived slights to national pride and sovereignty. And all the more so during their ten-year leadership transition cycle, which culminated in 2012 with the party's formal selection of Xi Jinping as successor to President Hu Jintao (Christensen, 2011). The timing of Tokyo's purchase and nationalisation of the Diaoyu/Senkakus was seriously ill-judged.

Since his installation and consolidation at the helm of the PRC government, President Xi Jinping has been pursuing the 'Chinese Dream', or what he also terms 'the Great Renaissance of the Chinese People'**.** The China that he took over 'had emerged from the global financial crisis cocky on the international stage' (Christensen, 2015, p. 28). This is a notably more assertive strategy then the one adopted by his predecessors: it is far removed from the dictum of *tao guang yang hui* ((韬光养晦) of former PRC Vice-Premier Deng Xaoping: hide brightness and foster obscurity (Lee and Shoon, 2015). To do so, President Xi has set in motion a proactive, two-pronged, foreign-policy initiative to strengthen ties with China's neighbours: one in terms of the Silk Road Economic Belt (an overland route) and the other in terms of the Maritime Silk Road of the 21st Century (a maritime route). Both routes leverage on China's economic strengths (Wong and Lye, 2014). This strategy also includes a US$50 billion-funded Asian Infrastructure Investment Bank (AIIB), created with the objective of lending money to developing countries to build roads, railways and other infrastructure, and subscribed to by 57 founding country members in 2015 – and creating a credible and lightly screened alternative to the Washington, DC-based World Bank/International Monetary Fund (Liu, 2015, p. 163). President Xi has committed China to a more activist global policy, including a more proactive role at the United Nations, whose mission of supporting peace-building and providing development assistance is shared by Beijing (Foot, 2014). Far from keeping a low profile, under Xi, PRC international diplomacy is emboldened, 'shifting from emphasising the role of serving the country's development to the role of serving the Chinese Dream' (Zhao and Gao, 2015, p. 35). Under Xi, China has rediscovered an assertiveness that claims a role for itself as a global force in a new era of major power relations. While seeking to bolster friendly relations with neighbouring countries, Xi has been resolute in maintaining China's assertive actions in dealing with territorial disputes with Japan to safeguard China's territorial integrity and sovereignty as well as its maritime rights. The PRC's annual military spending is now twice that of Japan's (Wong and Buckley, 2015). The new foreign-policy initiatives and assertive actions taken by Xi Jinping can be attributed not only to a recasting of Chinese diplomacy and international relations in line with the country's burgeoning economic muscle. They are also possibly feints to divert attention from economic malaise and social instability at home as well as political rivalry among different leadership factions (Sheng-Ping, 2014).

A re-pivoting by the US in East Asia: how to deal with the new China?

To this scenario of a rising, increasingly muscular, but also increasingly sensitive, China is a corresponding one of US waning. The twentieth century, the bloodiest on record, may have witnessed the start of 'the descent of the West' as much as 'a reorientation of the world' towards the East (Ferguson, 2006, n.p.). A rising China will necessarily alter, if not threaten, the preeminent geopolitical position that the US has enjoyed since the end of World War II, and more so since the end of the Cold War. The US response to China's steady rise in economic and political influence is possibly the single most notable event that will define international politics in the twenty-first century. How this response will unfold in practice remains as yet foggy. A continued engagement by the US (and other regional players, plus the European Union) with China is seen as an important strategy, meant also to bring that country more resolutely within the fold of the international community and its regimes, since nudging China to play the part of a responsible stakeholder and shaping its choices is, in itself, a recipe for stability and cooperation that reduces the risk of war. For example, China's support for, and participation in, UN peacekeeping operations – with some 2,200 Chinese nationals deployed in nine UN peacekeeping operations worldwide since the year 2000 – is a far cry from its rigid adherence to the doctrine of non-interventionism even a few decades ago (Zhang, 2015).

But the view from Beijing suggests a different narrative: a containment exercise is underway. In many disputes that involve China, the US tends to support the other party. The US supports the Philippines on the South China Sea dispute, and Manila may entertain re-inviting US forces back to their old naval base at Subic Bay (Hernández, 2015); the US has also agreed to sell more advanced weaponry to Taiwan. Meanwhile, the US labels China's overseas economic activities as neo-colonialist and considers China's territorial disputes with its neighbours to be evidence of expansionism (Jin, 2014). In April 2014, US President Obama reiterated Washington's security commitment to Japan, and – a first for a US President – he did not waffle (Ayumi, 2015, p. 78). Obama explicitly mentioned the Senkakus (and in their Japanese name only) as part of the US security remit, while dodging the sovereignty question:

> And let me reiterate that our treaty commitment to Japan's security is absolute, and Article 5 covers all territories under Japan's administration, including the Senkaku Islands . . . We don't take a position on final sovereignty determinations with respect to Senkakus, but historically they have been administered by Japan and we do not believe that they should be subject to change unilaterally. And what is a consistent part of the alliance is that the treaty covers all territories administered by Japan.
>
> (Obama, 2014, n.p.)

Although in a minority, other voices make the case for a different tack. Comparisons to 1914 and China's aspirations to graduate to some kind of regional hegemon are claimed to be overblown. Moreover, China's approach to its current expansion is that of an act that, rather than meant to be interpreted as belligerent, restores a regional prominence that the country has enjoyed for most of recent history (Shambaugh, 2013). Is the US prepared, in some measure, to recognise and accommodate to – which is not to capitulate to – a rising China by allowing it, or just watching it, assume what could be seen as its rightful position as the region's largest indigenous power (e.g. Gries, 2004; Kaplan, 2014; White, 2012)? Can a more sanguine, and less hawkish, US smoothen the transition from a *Pax Americana* to a *Pax Sinica* in the Pacific?

In any case, China's neighbours are nervous about such a policy. They have gravitated towards even stronger security relations with the US in response to China's heavier and longer shadow. Cases in point are the 'enhanced defence cooperation agreement' signed between the Philippines and the US in 2014 (Eilperin, 2014); and a 25-year 'force posture' agreement signed between Australia and the US in the same year (Murphy, K., 2014). Beijing may have started to realise that its more assertive and coercive actions towards its neighbours are causing these to seek to deepen and consolidate further their defence relationship with the US (Christensen, 2015, p. 298). Also because of the volatile regime in North Korea, and its fractious relations with its neighbours, the latter fear an abdication of US stewardship in the Pacific, creating a dangerous power vacuum, let alone Beijing's usurpation of that hegemonic role. They are distressed that the re-pivoting or re-balancing by the US – under congressional pressure to slow military spending – in the Asia-Pacific may amount to none other than disengagement (Giacomo, 2014; Lieberthal, 2011; Logan, 2013). There have been calls in the US for its forces to pull out, leaving Japan to defend itself (e.g. Bandow, 2014). Japan, in particular, has been most vociferous in seeking assurances from Washington that any such US extrication will not happen (Ding, 2013); hence Obama's statement, reproduced above, offering reassurance. (The commitment has since been confirmed by US President Trump's Secretary of Defense, Jim Mattis.)

Bringing back Japan?

Japan has not been sitting on the fence during all this. The US and other East Asian nations, including China, have their own reasons to feel anxious with changes under way in Japan. In a clear reference to Tokyo, a senior official of the Chinese People's Liberation Army General Staff deemed fit to state that China would never allow 'ruthless, fascist and militaristic aggression to stage a comeback' (Wang, 2014). The reference is intended to critique the policies of the Japanese government that is cementing its US alliance, while forging strong economic and political relations with East Asian countries. More seriously, the re-elections of Shinzo Abe of the Liberal Democratic Party (LDP), as Prime Minister of Japan in 2012 and again in 2014, have ushered in a more *gung ho* Japan. Evidence of this includes the implementation of the so-called 'three arrows' of Abe's (rather left-wing) economic policy, also dubbed 'Abenomics':

fiscal spending, monetary easing and a growth strategy. But what also came along was a clear shift towards a more nationalist and regionally assertive Japan; and the developed world's most right-wing government (Murphy, R., 2014, p. xii). Abe is an avowed revisionist, keen to extract the country out of a pacifist constitutional morass that, he alleges, had been hoisted by the US on an occupied and humiliated Japan in 1947. The Japan–US alliance has been shored up, and partnerships and cooperation agreements with many East Asian countries strengthened; with regional suspicions of China on the rise, the dictum that 'the enemy of my enemy is my friend' speaks volumes. Japan has formally lifted its decades-old ban on the export of weapons, and opted to surge its military spending to unprecedented levels (Fackler, 2015b; Panda, 2015). It has revised textbook-screening guidelines to give Japanese children a more patriotic twist to modern Japanese history and to better reflect the government's official position on territorial issues (Cai, 2014). The Abe government in Tokyo has also set up a Japanese National Security Council, announced the first National Security Strategy and established the National Defence Programme Guidelines that introduced the concept of a 'dynamic Joint Defence Force'. This emphasises the Self-Defence Forces' joint operations and interoperable capability at sea, in the air and on land, and help bolster Japan's defensive bulwark in the southwest, and in particular deploy some 600 well-armed troops in the Nansei island chain that includes Okinawa and the disputed Diaoyu/Senkaku islands (Keck, 2015; Takahashi, 2014). Two helicopter anti-submarine carriers, the *Hyūga* and *Ise*, have been joined by the larger *Izumo* in 2015, and one other carrier will follow by 2017 (Gady, 2015). The rising tensions over the Diaoyu/Senkaku have also been useful in Washington's efforts to 'urge Tokyo to assume a larger role in its alliance with the US' (Christensen, 2015b, p. 32). In July 2015, the Diet (the lower house of Japan's Parliament) approved two controversial bills that would change the country's security laws, despite large public protests: the changes would allow Japanese troops to fight overseas for the first time since WWII (BBC News, 2015). Might it prove a Pyrrhic victory for Abe? Various experts have said that the bills are unconstitutional, and negative public opinion had been riding high (*The Economist*, 2015e). Nevertheless, final passage by Parliament of the legislation authorising overseas combat missions for Japan's military, overturning a decades-old policy of reserving the use of force for self-defence, was secured in September 2015 (Soble, 2015).

This situation helps explain why a deeply pacifist Japanese public has been drawn into a zealous nationalist and patriotic stance. The Diaoyu/Senkaku affair has become one point of tension that has been used to expose the potential consequences of Japanese indifference to China's growing bluster as a regional player. It has provided Abe with the justification he has needed to drive his militarist agenda in the face of an overall sceptical national constituency. Yet, in the 'chicken and egg' scenario that bedevils international relations, the Japanese tack towards pugnaciousness is dangerous, and legitimates China's own heightened response (*New York Times* Editorial, 2014). Is this the vicious spiral that could yet drive these two countries, the region and the US to confrontation? Will the

Diaoyu/Senkakus be to Shinzo Abe what the Falklands/Malvinas were to UK Prime Minister Margaret Thatcher in the early 1980s?

In Abe's own words, he is seeking a 'departure from the post-war regime' by 'bringing back Japan' (Watanabe and McConnell, 2015, p. 14). Although Abe has never said from what, and to what, he will bring back the nation, many Japanese believe that what he means is to bring back a militarily, diplomatically and economically strong Japan from the political nadir, economic abyss and humiliation of the immediate post-WWII decades. Locked in a special alliance with such a Japan, the US would be dismayed, uncomfortable, even appalled by such a dangerous embrace (Halpin, 2015).

Conclusion

In the 2005 movie and epic historical drama *Kingdom of Heaven*, the Crusader knight Balian ends up leading the ill-fated defence of the Holy City of Jerusalem against the ascendant Arabs. While negotiating terms of surrender, Balian asks the Arab Sultan Saladin: 'What is Jerusalem worth?' Saladin first replies 'Nothing' and walks away. But soon he stops, turns and adds 'Everything' (*Kingdom of Heaven*, 2005). Even after a thousand years, peace in Jerusalem and the Holy Land remains elusive.

The Diaoyu/Senkaku islands may carry an uncannily similar destiny. They are a set of insignificant and uninhabited rocks and islets: in themselves, worth nothing. But, having assumed the status of 'inherent and indisputable sovereign territory' for both China and Japan, they may now be worth considerably more. They find themselves at the very centre of a storm that has multiple origins, including a rising and increasingly self-confident China, a wary yet progressively nationalist Japan and a US looking for a suitable makeover. Like Jerusalem, might a lasting *denouement* to the Diaoyu/Senkaku spat remain intangible, even for another millennium?

In the shadow of the 100th anniversary of World War One (WWI) (1914–1918), some sober reflections about the significance of East Asian politics for world peace and stability are quite in order. Mackinder may be right again: any changes in that region, instigated by and impacting on, both China and Japan, but impacting on other players, may be deeply and lastingly significant to international relations. Back in 1995–1996, it was tensions in the Taiwan Strait that were brewing and threatening conflict: then, the contending parties were the PRC and Taiwan/Republic of China (ROC), with Beijing conducting war games, live firing exercises and missile tests aimed clearly at Taipei; while two US aircraft carriers patrolled nervously at hand (Scobell, 2000). Now, the focus may have shifted, but only by a few hundred kilometres, to the Diaoyu/Senkaku chain.

The churn in the Asia Pacific region is palpable. The next chapter will look more closely at the two key protagonists embroiled in this latest dispute in East Asia: China and Japan. By looking at their past and the key role of 'historic memory' in each, the chapter hopes to help unravel their national contemporary psychology, and the twists and turns that impact on their policy standpoints with regards the Diaoyu/Senkaku islands. Notably, the chapter also looks closely at the 'wild cards' that distort and complicate the respective position of either side: Okinawa/Ryukyu, in the case of Japan; and Taiwan/ ROC, in the case of the PRC. In recognising that problems may be optimistically re-read and re-interpreted as opportunities, Chapter 3 will seek to explain how Okinawa and Taiwan bring their own added value and insights to a potential solution of the Diaoyu/ Senkaku affair.

3 The China–Japan relationship through the ages

> Whereas love of home draws on good memories, love of nation-state draws on not only good memories but also on bad ones of defeat and humiliation, unhealed wounds that easily metastasize into hatred of the unassimilated inside and aliens outside the borders of homeland.
>
> (Yi Fu Tuan, 2014)

Introduction

It was obviously an awkward moment. Shinzo Abe met Xi Jinping for the first time on 10 November 2014, when the latter was hosting the Asia-Pacific Economic Cooperation (APEC) summit in Beijing. A public handshake could hardly be avoided, but the frostiness in the encounter was tangible. Abe and Xi are leaders that share certain qualities and political ambitions: ensconced at the helm of their respective governments, they exude a keen desire to project their country as a respected regional power; brandish an agenda for economic growth and 'soft power' political influence; have each approved the largest military budgets ever in their respective country's history; affirm a need to restore pride and dignity to country and nation; and hope to be able to execute a reversal of the tragic events of the recent past. Some sense of their nationalist credentials can be gleaned by both countries' refusal to recognise dual citizenship. The two countries have also fought two major wars against each other in the last 120 years: 1894–1895 and 1937–1945.

Talk of 'made in China' today no longer conjures up sniggering comments about low-quality copies, and 'created in China' is a growing economic reality (Keane, 2006). Meanwhile, all the hubris about Japan being the economic dynamo 'in the passing lane' (Kamata, 1983) is long gone; the Japanese economy has weakened and – given its strong resistance to immigration – the Japanese population is both contracting and aging. Thus, China's rise coincides with its neighbour Japan's decline as the leading Asian economy. One country's expansionism is the other's country's nightmare. To some observers, it is just a question of time before the political tectonic plates collide.

This chapter reviews the troubled rapport between China and Japan throughout the ages. One of the pivotal mistakes in seeking to critically understand China and

Japan today is to analyse them artlessly using a Western-driven perspective. A greater appreciation of events in East Asia might help to introduce an element of humility into an otherwise glib superiority complex. A better sense of local geography and history will also help put matters in a more nuanced perspective. Thus, allow me some indulgence to venture into a Braudel-style *longue durée* mode of writing that gives priority to long-term historical dynamics (Braudel, 1958). After all, context is everything. And, in East Asia, lingering history and representations of the past can and do continue to stymie cross-country collaboration and influence policy decisions.

Also presented here are the related but distinct positionalities of both Taiwan/Republic of China and Ryukyu/Okinawa: looking at the Diaoyu/Senkaku islands from their perspective adds new colour and drama, as well as opportunities for creative insight, to the festering affair. Moreover, in agreement with Hara (2012), and as mooted by both Katayanagi and Iwashita in this volume, a four-player approach to the Diaoyu/Senkaku affair appears to hold more promising possibilities for solution than a strictly two-player one. Finally, it is pertinent to note that Taiwan remains that part of China that is most nostalgic about, and sympathetic to, Japan and its culture: note the recent craze in Taiwan for Hello Kitty 'Made in Japan' merchandise (*The Economist*, 2016b). Also, Okinawa/Ryukyu is that part of Japan that is most suspicious of, and angry at, Tokyo, while also having the strongest and most consistent historical ties to Taiwan and China.

But, first, an important proviso: the entities that we now call 'China' and 'Japan' are creatures of the modern age. We assume that we are dealing with sovereign states and which have clearly marked territorial borders, and within which they exercise undisputed authority. No wonder they are keen to stake claims to islands that would defend or extend their territorial lands (and waters). UNCLOS has extended this paradigm over large swathes of ocean. And yet, claims to the Diaoyu/Senkaku islands can be traced back to times when such demarcations were not common: in pre-modern Asia – which, where China and Japan are concerned, was certainly the case until late in the nineteenth century – it was geographical *centres* such as Edo/Tokyo, Kyoto, Shuri, Fuzhou and Beijing that were clear; while *borders* and borderlands were not (Smits, 2015, p. 10). Under the logic of this pre-modern system, rulers projected their power beyond borders by their 'benevolence' or 'virtuous' rule, which – as Confucian thinkers believed – would lead foreign powers to acknowledge and acquiesce to the ruler's moral suzerainty. Thus, it was possible to rule even where there was no mechanism of physical governance in place (Lee, 2011, p. 5).

Thus, the extent of pre/early-modern Japan's physical territory was quite similar to the modern-day state. However, the boundaries of this territory were 'fuzzy': in the north, Japan 'faded' into Ainu land in Hokkaido, which in turn faded into Russia; in the south, Japan faded into the Satsuma kingdom (based in Kyushu), which faded into Ryukyu, which in turn faded into the Chinese kingdom. Cultural boundaries did not tally with political ones, since the latter were so tenuous (Smits, 2015, p. 10; Walker, 2015). No wonder, therefore, that 'To whom did the Diaoyu/Senkaku islands belong?' was a question that simply did not arise.

中国: China, the Central Kingdom

One seminal event of the so-called 'Age of Discovery', according to the Western canon, was the first voyage of Christopher Columbus, a Genoese sailor in the employ of the King and Queen of Spain, in 1492. Columbus sailed with a complement of 90 sailors on three ships, the biggest of which was 26 metres (85 feet) long. And yet, contrast this to what happened some 70 years earlier: between 1405 and 1433, during the reign of the Yongle Emperor Zhu Di (朱 棣), the third emperor of the Ming Dynasty, Admiral Zheng He led seven major naval expeditions, commanding the largest armada the world would see for the next 500 years. Zheng He's fleet included 28,000 sailors on 300 ships; the largest vessels in this formidable armada were each about 122 metres (400 feet) long. These ships also incorporated advanced design elements that would not be introduced into Europe for another 350 years, including balanced rudders and watertight bulwark compartments (Kristof, 1999).

It may have just one time zone, but China is huge. It currently has the world's largest population: at 1.3 billion people, just under one out of every five living persons in the world today is a Chinese national. It is the third largest country in the world by land area (after Russia and Canada). It is one of the five permanent members of the UN Security Council. It is (along with Russia) the country with the largest number of land borders: with no less than 14 different countries sharing a 22,000 km long boundary (Shen, 2012). It is the world's largest economy by purchasing power parity. Written records relating to the history of China have been dated back to 1200 BC. No one should be surprised that China has called itself 'the Central (or Middle) Kingdom'. It is simply, by definition, the country at the centre of the world.

Unlike Japan, China is essentially a continental state, and largely continues to think and act like one. Its survival has depended on stemming invasions from its long western and north-western land frontier; hence the iconic Great Wall of China. China has historically dealt with its security environment by developing and maintaining a series of expanding, concentric rings of security around its territory, especially the Han heartland. However, while its northern and western frontiers were thus protected, the 'arc of security' did not incorporate and extend to China's coastal southern and eastern flanks, leaving the country vulnerable to infiltrations by European states, with their technologically advanced sea power. Thus, throughout its recent history, China's security and sovereignty were only severely compromised by its failure to develop maritime power sufficient to contain or overcome Western naval technologies (Dutton, 2014, pp. 7–8). More about this below.

China's inherent continentalism offers more insights. Continental powers, like China, tend to think about (and of) the sea differently than more 'natural' seafaring states like the Netherlands, the Philippines, Britain (and Japan?). Where nautical peoples see the oceanic expanse as a commons – an ungoverned space, open for the free use of all – continental peoples see national territory, to be governed as though it were dry land. This attitude might explain China's readiness to treat the South China Sea – and not just the islands, shoals and reefs within it – as Chinese territory and subject to Chinese domestic law (Holmes, 2015).

Around 40 million citizens of China live on islands: 23 million of these on Taiwan; 9 million on Hainan; and 7 million more on Hong Kong. That makes less than 3 per cent of the total resident population. China has settled various border disputes, including *all* those involving land borders that had been pending with Russia and the post-Soviet neighbouring states of Kazakhstan, Kyrgyzstan and Tajikistan (Blank, 2011; Taylor Fravel, 2008). Apart from the Diaoyu/Senkaku case, China has a dozen other unresolved territorial disputes, including a fairly complicated situation along its border with India, and which led to war in 1962 (Shah, 2015). Where islands are concerned, these border disputes include: another, differently complex situation in the South China Sea (Samuels, 2013); the island of Shindo (in Korean) or Shin (in Chinese), currently occupied by North Korea, along the China–North Korean border (Pinilla, 2004); as well as a disputed reef (Socotra Rock in English; Ieodo in Korean; Suyan Rock in Chinese), a submerged feature, that is currently controlled by South Korea in the East China Sea but about which Beijing has been increasingly vocal (KBS World, 2012). The Ieodo/Suyan dispute is officially dormant, subject to a 1978 bilateral agreement that will remain in force for 50 years.

For almost two thousand years, and particularly during the Ming (1368–1644) and Qing (1644–1912) dynasties, China maintained an imperial tributary system of trade networks and foreign relations with various regional powers. The latter included Korea, the Ryūkyū Kingdom (comparable to present-day Okinawa), Annam (Vietnam), Siam (Thailand), Burma (Myanmar), Nepal and even Japan (Kang, 2010). Unlike other tribute systems around the world, the Chinese version consisted of mutually beneficial economic arrangements: tributary states maintained their political autonomy while recognising the suzerainty of China. The tribute system facilitated frequent economic and cultural exchange: thus, the various dynasties of Imperial China deeply influenced the culture of peripheral countries and drew them into a China-centred, or Sino-centric, international order. Stability was predicated on hierarchy and tribute, not on any notion of a balance of power (Kang, 2005); and China was the obvious and uncontested power at the apex of the chain of command. It could not be otherwise. Even when Western powers started appearing and requesting permission to trade with China, a tributary relationship was naturally assumed by the Chinese.

But this natural order was to be shaken to its core. By the 1830s, Chinese exports (including silk, ceramics and tea) were in high demand in Europe (particularly the United Kingdom) and a heavy trade deficit ensued. For example, already by 1800, the British East India Company was buying 23 million pounds of tea (just over 10 million kilos) per year, at a cost of £3.6 million of silver (Perdue, 2011). Opium, it seemed, was the only commodity that could save the balance of payments in Britain from ruin. And it did: indiscriminate flooding of opium from India into China by European merchants, often with the connivance of local middlemen, robbed the Qing dynasty of the revenue that had been flowing into its coffers for centuries. But there was more. Opium was not a harmless commodity: it was an addictive substance with noxious, and sometimes tragic, consequences for its users and their families (Yangwen, 2005); it also eroded self-esteem, shamed the nation and cast a damaging pall on the reputation of the Chinese people, while it propelled the relative supremacy of the (non-addicted) Western coloniser (Brook and Wakabayashi, 2000, pp. 3–4). When the Chinese government sought to rein in this damaging practice, and destroyed significant opium stocks in Canton, the Western Powers declared war, citing the principles of free trade, commerce and 'civilisation', first in 1839–1842 and then again in 1856–1860 (Fairbank, 1953). These were China's first armed conflicts with the West, and a critical juncture in its contemporary self-image. They constitute a national wound: the start of a Western conspiracy to destroy China with drugs and gunboats. China's national myth as emergent from the Opium Wars continues to mould its interactions with the outside world, and shapes both its diplomatic and popular responses to world events (Wong and Twitchett, 2002). By textbook convention, the Opium Wars constitute the moment of awakening, the beginning of the inevitable historical process leading to the victory of the Communist Party in 1949, a hundred years later. Public historic memory is spun to serve the present; and delusion and prejudice continue to bedevil China's relationship with the contemporary West, and vice versa. The Opium Wars are the fuel for a powerful narrative of national victimhood, reanimated at various times to unite the Chinese public against outside threat; as well as to divert attention from the state's own shortcomings. Cronyism, bureaucratic bungling, deep corruption and internal (especially interethnic) conflict facilitated the nineteenth-century victories of the West against China (Gelber, 2004; Lovell, 2011).

Via the Unequal Treaty of Nanking (1842), China had to pay Western powers reparations and costs of US$21 million, condone the opening of five more treaty ports for the purposes of trade and cede the island of Hong Kong to Britain. Via the Unequal Treaty of Tianjin (1858/1860), China had to make further heavy payments, cede Kowloon and parts of Outer Manchuria, and allow the importation of opium. By the end of these Opium Wars, China had to also accept the diplomatic presence of Britain, France and Russia in Beijing: a clear indication that the long-standing tributary system had well-nigh collapsed. It was a thorough defeat and utter humiliation of China by the West (Hsü, 2000).

Some sense of self-deprecation followed. Rather than dwelling on anger, Chinese writers in the nineteenth century used the memory of the Opium Wars as a spur toward modernisation, seeking answers as to why Imperial China was so weak in relation to the Western powers (who won the Opium Wars easily in spite of having serious logistical challenges) (Lovell, 2011). It was not until General Chiang Kai-Shek's Nationalist regime in the 1920s and 1930s that school textbooks began their revisionist project, teaching about 1839 and its aftermath as the bedrock of national shame, the imperialist humiliation that all Chinese citizens should work together to avenge. The Communist Party of the PRC can be seen to be continuing such a quest of retribution today. After all, it was as recently as 1997 and 1999, that the last vestiges of colonial rule were removed from China, with the return of Hong Kong and Macau to the PRC, by the United Kingdom and Portugal respectively.

Meanwhile, as if they were not bad enough, the Opium Wars were just the beginning of a century of grave humiliation for China. Its neighbour Japan was on an expansionist and modernising wave, keen on securing access to raw material, food and labour reserves to power its economic development. The First Sino-Japanese War (1894–1895) led to China's loss of Korea, Taiwan, the Pescadores (Penghu) islands and the Liaodong Peninsula to Japan, ratified by the Treaty of Shimonoseki (1895). (This, by the way, is the same treaty that Beijing interprets today as having also transferred the Diaoyu/Senkaku islands to Japan, even though these islands are not explicitly mentioned in the treaty.) This victory established Japan as the dominant power in Asia. Meanwhile, on the Shandong peninsula, the port city of Qingdao was leased to Germany (1897) and the port city of Weihai to Britain (1898). Then, in the aftermath of the Boxer Rebellion, the 1901 Boxer Protocol imposed further indemnity payments on China, and obliged the Qing government to allow eight foreign countries to base their troops in Beijing. After all this mortification, and a general frustration at the slow pace of reform, the situation led to the Xinhai Revolution in 1912, which ended 2,000 years of dynastic rule in China. But this era of republican rule was short-lived: shifting coalitions of competing provincial military leaders controlled the country after 1916. Still more humiliation and public anger were in store with the Treaty of Versailles (1919) that brought an end to World War One – and which China refused to sign – and which granted former German holdings in East Asia and the Pacific – including the Shandong peninsula – to China's regional nemesis, Japan. (Shandong reverted to Chinese control in 1922, after mediation by the United States.) As a fresh wave of modernisation gripped the country, a bitter civil war ensued between the Kuomintang (Nationalist Party or KMT) under Chiang Kai-Shek and the Communist Party of China (CPC) under Mao Tse-Tung. This continued, openly or covertly, through the 14-year-long Japanese occupation of various parts of the country (1931–1945), and what has been termed the Second Sino-Japanese War. Japanese forces committed numerous war atrocities against the Chinese civilian population at this time (but more about this below). After the end of World War Two in 1945, the civil war in mainland China continued until the Nationalist government forces were defeated by the CPC forces in 1949. At that point, the Nationalist government retreated to Taiwan and other smaller islands with its remaining forces, along with most of the KMT leadership and sympathisers. At the helm of the Republic of China, the Taiwan-based government continued to represent China in international relations for almost two decades. Taiwan/ROC kept representing China as a member of the United Nations (UN) – and including the UN Security Council as one of its five permanent members – until 25 October 1971, when it was expelled, and its UN seat was transferred to the PRC.

Meanwhile, the tide had started to turn earlier, in favour of China and against Japan. At the Cairo Conference of November 1943, the Allied powers of WWII (China included) agreed to punish the aggression of Japan by restoring *all* the territories that Japan had annexed in China, including Manchuria, Taiwan and the Pescadores, to China, and to expel Japan from the Korean Peninsula. (This commitment towards restoration, according to Beijing, should also apply to the

Diaoyu/Senkaku islands.) This decision was confirmed by the signing of the Japanese Instrument of Surrender of 1945.

Napoleon Bonaparte is credited with describing China as a sleeping giant that should be left unmolested for, if and when it wakes, it will move the world (Aliberti, 2015, p. 185). Well, it appears that the giant *is* waking up: the twenty-first century finds an upbeat China that, domestically, has its communist elite upholding its mission as the defender and expression of the Chinese nation; and, internationally, is positioning itself once again as a great power in the international order, having closed the chapter on its (100-year) historical hiatus as a victim of Western humiliation (Harnisch, 2016). The few items of unfinished business from this traumatic bygone epoch include Taiwan/ROC with its islands (Baldacchino and Tsai, 2014); and of course the Diaoyu islands, yet to be returned and reunited with the motherland.

Wild card: Taiwan/Republic of China

Having an ISO two-letter country code (TW), a UN/ISO 3166-1, three-digit numerical code (158) and an extant Internet Assigned Numbers Authority (IANA)-assigned country code top-level domain (.tw) would suggest that Taiwan is another of the world's sovereign states. And 21 member states of the UN (along with the Holy See) *do* recognize Taiwan as such. But the other 172 do not.

When the Kuomintang escaped mainland China and set up its base in Taiwan in 1949, it also carried the flag and seat of government. It effectively became the government of the Republic of China (ROC) in exile. This means that the Taiwan government traces its political lineage to the ROC set up after the revolution in 1911 in China (and whose flag it continues to fly). For many years, KMT Generalissimo Chiang Kai-Shek waited for the right opportunity to return to mainland China and re-establish the seat of government in Beijing. But the situation was soon reversed: it was Taipei that became fearful of a mainland Chinese invasion that would bring Taiwan under the rule of the Communist Party, now resolutely running the PRC from Beijing. Moreover, with the increasing economic and political clout of the People's Republic of China, allegiances started to waver and switch sides. After October 1971, Taiwan/ROC was expelled from the United Nations, and the PRC was recognised as the sole lawful representative government of China.

This means that, *de facto*, jurisdiction-wise, two Chinas have existed since 1949. Nevertheless, the PRC has instituted a 'One-China policy' under which it actively opposes any moves seen to endorse the ROC as an independent sovereign state. Officially, the view from Beijing is that Taiwan and its islands (mainly the Penghu, Kinmen and Matsu groups, but also others, including the uninhabited Diaoyu chain) constitute a 'rogue province' of China. Respectful of Taiwan's particular recent history and turbulent relationship with mainland China/PRC, and knowledgeable of Taiwan's wariness of the PRC's ambitions and motives, Beijing has offered Taiwan the application of the 'One China, two systems' creed and therefore the status of a 'special administrative region' (SAR). This arrangement is similar to that enjoyed by Hong Kong since 1997 and Macau since 1999, although reunification with the motherland remains the long-term objective (Jiang, 2014). While Taiwan has not obliged, Beijing has put in place various domestic administrative structures that already treat Taiwan as an SAR (e.g. Ministry of Foreign Affairs of the PRC, 2015).

Events in the Taiwan Strait have tended to confirm Taiwan's hesitation. First, the botched attempt at the military invasion of Taiwan via Kinmen island by the PRC in 1949; then, Kinmen was the target of indiscriminate shelling from 1954–55 and again in 1958, followed by more on-off bombardment. The last tense confrontation was in 1996 when China carried out military exercises and missile tests in the Taiwan Strait, while the US deployed two aircraft carriers in the same region (Ross, 2002). At that time, the Taiwan Strait was seen as the most likely trigger to the next world war (Christensen, 2002); just as the Diaoyu/Senkaku affair suggested, a decade later.

It was only in 1987 that the Taiwan/ROC government began to allow visits to mainland China. The demise of authoritarian KMT rule in Taiwan after 1992 ushered in a shift away from Taipei's own 'One-China' policy – that saw all China reunited under the ROC government – and, to Beijing's consternation, an independence movement emerged, championed largely by the Democratic People's Party (DPP). During the earlier period of DPP rule (2000–2008), the PRC reacted by seeking the diplomatic isolation of Taiwan, and the passage of an Anti-Secession Law in 2005 formalised 'non-peaceful means' as an option of response by Beijing to a formal declaration of independence in Taiwan.

Since 1979, the US recognises the PRC government as the sole legitimate government of China, acknowledging the Beijing position that there is only one China and Taiwan is part of China. Thus, the US does not support Taiwan independence. However, the 1979 Taiwan Relations Act provides the legal basis for an 'unofficial' US–Taiwan relationship, in the absence of a diplomatic relationship or a defence treaty. It codifies the US's ongoing commitment to assist Taiwan in maintaining its defence capability. The US insists on the peaceful resolution of cross-Strait differences, opposes unilateral changes to the status quo by either side and encourages dialogue to help advance such an outcome; but, at the same time, it does not shirk from negotiating the sale of high-grade military technology to Taipei, in spite of Beijing's condemnation (Reuters, 2014). Beijing argues that the Taiwan question has been the most serious irritant in its relations with the US. It insists that US arms sales to Taipei infringe on China's sovereignty because, after all, Washington acknowledges that Taiwan is part of China. The prevailing discourse in the US remains, however, that allowing the PRC to achieve the military upper hand in the Taiwan Strait is tantamount to abandoning Taiwan to a forced unification. Such an act might considerably enhance China's military and geostrategic position in Asia, at the detriment of both the US and its allies (Rehman, 2014).

The long-strained relations between Beijing and Taipei have significantly eased since 2008. With the KMT back in power through the Presidency of Ma Ying-jeou (2008–2016), dialogue through semi-official organisations was resumed.

Cross-strait relations, particularly in terms of the 'three links' (transport, commerce and communication) have improved markedly: direct flights commenced between the two sides; and Taiwan started allowing a daily quota of up to 3,000 visitors from mainland China (Kuo, 2015, p. 9). The first government-to-government contact (at minister level) between the two sides took place in 2014 (Wang, 2014). Other examples of the increased cooperation across the Taiwan Strait include China's approval for Taiwan to secure observer status at the UN's World Health Assembly in 2009; and the tacit agreement that neither side will poach each other's existing diplomatic partners. In the PRC, the possibility of inviting Taiwanese companies to join its seabed oil/gas exploration and exploitation in the East China Sea has also been raised (Guo, 2010).

In January 2016, Tsai Ing-wen convincingly won the presidential elections in Taiwan as the DPP candidate. Beijing is again looking nervously across the Taiwan Straits, while Tsai seeks closer ties with countries other than China and seeks to wean the island off what had steadily become a heavy economic dependence on its large neighbour. Other symbolic steps also suggest she is reversing her predecessor's pro-Beijing push and not acknowledging the 'One-China policy' (Ting, 2016).

Meanwhile, Japan has not been sitting on the fence. It has sensed, and exploited, an opportunity to play off Beijing and Taipei against each other with regards to at least one feature of the festering Diaoyu/Senkaku affair.

Beijing and Taipei may disagree on many issues; but one thing on which there is a practical unanimity of opinion has to do with claims to the East China Sea. Their position on the Diaoyu/Diaoyutai as inherent territory of China is identical; and their protests to the US in 1971 (prior to the Okinawa reversion agreement coming into force in 1972) – by the ROC in February and June; and by the PRC in December of the same year – to return the Diaoyu islands to China were couched in similar language, though issued independently (Selden and Susumu, 2014).

However, a clear difference in tone between Beijing and Taipei about the contested island group was evident post-2012. While Beijing's tone grew more strident, that in Taipei sounded softer. A new 'East China Sea Peace Initiative' (ECSPI) was proposed by Taipei, whereby the ROC government reaffirmed its position of 'safeguarding sovereignty, shelving disputes, pursuing peace and reciprocity, and promoting joint exploration and development' (ECSPI, 2012). (Taiwan's former President Ma Ying-jeou has since (in May 2015) unveiled a similar *South* China Sea Peace Initiative, calling on all concerned parties to 'exercise restraint, safeguard peace and stability in the region, and refrain from taking any direct action that might escalate tensions' (Coutaz, 2015)). Taipei has sought to play the opportunistic honest broker here, asking the PRC and Japan (but also pledging its own part) to 'refrain from taking any antagonistic actions; shelve controversies and not abandon dialogue; observe international law and resolve disputes through peaceful means; seek consensus on a code of conduct in the East China Sea; and establish a mechanism for cooperation on exploring and developing resources' in the same sea (ECSPI, 2012). One must remember that the Taiwanese are amongst the most sympathetic to Japan in East Asia, with an elderly generation nostalgic towards Japan and the island's occupation by that power until

1945, and a younger generation charmed by Japanese manga and anime; indeed, Taiwan was the top donor to Japan's 3/11 earthquake recovery fund (Chen and Hwang, 2015).

Moreover, Taipei clinched a non-official agreement in 2013 with Tokyo that secured opportunities for Taiwanese fishers to fish in the waters around the Diaoyutai/Senkaku, beyond the 19 km (12 mile) 'territorial' sea (Chinese, Taiwanese, Japanese? – no matter) surrounding the islands. The deal gave Taiwan access to an additional fishing zone of 4,800km^2 (1,400 square nautical miles) outside Taiwan's temporary enforcement line. Under the deal, fishing vessels from both countries can now operate in a large area within the designated zone without being subject to the jurisdiction of the other side, while a smaller area of the zone, where Japanese fishing vessels frequently operate, is under joint management by the two governments. In a nod to Beijing, the accord states that it does not have any bearing on sovereignty claims by any country to the islands (Shih, 2013). Both Taiwan and Tokyo have scored diplomatic points with this agreement, showing that 'it is indeed possible for Asian powers to shelve big disputes indefinitely while getting on with everyday life' (Holmes, 2013, n.p.). For Japan, it has also succeeded in placing 'a wedge between Taiwan and China, preventing them from forming a united front against Japan's interests' (Kawashima, 2013, n.p.). Amendments in 2015 include the proviso that Taiwanese and Japanese fishing boats take turns operating in the area north of the Yaeyama islands and maintain a 7.4 km (four nautical mile) distance between each boat while fishing there (*Want China Times*, 2015). This accord is also important for the arguments proposed in this text, because it embraces some of the 'solution protocols' that are proposed in these pages, and which have been applied, fairly successfully to other island disputes. Their implementation to at least one component of the Diaoyu/Senkaku affair is quite telling.

The chapter now turns to look at the other key stakeholder in the Diaoyu/Senkaku matter, Japan, but also discussing the wild card that complicates the Japanese position, but may yet also offer a window of opportunity for dispute resolution: the Ryukyu Islands, or modern-day Okinawa, the only island prefecture of Japan.

日本: Japan, Land of the Rising Sun

Japan's millenary relationship with China is evident even from its very name. Nippon, the country's name in the vernacular, or its more locally common variant Nihon (にほん), literally means the place where the sun originates from, or the east: this refers to Japan's eastern position relative to – you guessed it: China.

Japan has *no* land borders with any other country: it is a wholly island nation, one of the world's only four island states with large populations (along with Indonesia, Philippines and the United Kingdom). It is a volcanic archipelago of no less than 6,852 islands, of which 314 are inhabited, located on a rough north–south axis (Japan Statistical Yearbook, 2015). Since its borders were reconstituted after WWII, Japan has had – *de facto* – only four border disputes (and of which only one is officially recognised as such by both sides: see below). Yet, Tokyo has not managed to resolve any of them. A combination of island geography and significant geographical distance from neighbouring countries (Korea, China, Russia) has provided Japan with a natural isolation and protection from the outside world for most of its pre-modern history (Pyle, 2006, p. 397). Like other totally islanded states (and unlike China), in the case of Japan, *all* things new, including all invasions, come from across the water. Nevertheless, to the Japanese, the four main islands of (from north to south) Hokkaido, Honshu, Shikoku and Kyushu are considered to be *naichi*, the Japanese mainland – (内地); literally, 'inner lands' – and are interconnected by bridges and tunnels. Honshu is the world's second most populated island (after Java) with over 100 million residents. At 4,480,000km^2, Japan is the country with the world's ninth largest EEZ. (For comparison's sake, China comes in as 19th, with an EEZ one fifth the size of Japan's.)

Apart from the Diaoyu/Senkaku case, Japan has three other current territorial disputes, but all involving islands. None of these three disputed island territories is occupied or administered by Japan (Kapur, 2012; Stephan, 1974).

One dispute deals with the Liancourt Rocks: two islets and surrounding small rocky outcrops with a combined land area of 18.8 hectares (46.3 acres), known as Dokdo (in Korea) and Takeshima (in Japan). Since 1965, Japan has protested that South Korea has unlawfully occupied these islets; while South Korea responds that there is no dispute to be settled with Japan regarding their sovereign status (Bong, 2013; Flamm, 2014; Weinstein, 2006).

A second dispute, this time with Russia, concerns the larger Southern Kurile islands (known in Japan as the Northern Territories). The four disputed islands comprise a land area of almost 5,000 km^2, and are thus by far the largest implicated in Japan's ongoing border issues. These islands were Japanese for seventy years (1875–1945) until they were occupied by invading Soviet forces in the dying days of WWII, and their Japanese residents forcibly evicted. Strangely enough, since the collapse of the USSR, it is the Japanese state that has become the main provider of welfare, infrastructure and economic opportunities for the current Kuril islanders. Because of this lingering dispute, Russia and Japan have not yet signed a peace treaty to end WWII (Burrett, 2014; Richardson, 2016).

A third and (rarely mentioned) dispute, also with Korea, concerns Socotra Rock or Ieodo (to the Koreans) and Suyan Rock to the Japanese: a submerged rock with a helicopter pad and a Korean research station, and over which China has also lodged claims. To overcome deadlock in maritime boundary delimitation negotiations because of Ieodo, Japan reached an agreement with Korea in 1974 on joint development with respect to the broad area (29,092 square nautical miles) encompassed by their overlapping claims to maritime jurisdiction in the East China Sea (Agreement between Japan and Korea, 1974). The Japan–Korea joint development zone shelved the issue of boundary delimitation (see Chapter 11) and is designed to facilitate the exploration for and exploitation of seabed oil and gas resources over a 50 year period although, to date, without success (Kim, 2012).

The first historical documents mentioning Japan date to around the fifth century. Jimmu, the first emperor of Japan, was installed in around 600 BC; the country has maintained that imperial tradition to this day, and is the oldest of 27 extant monarchies around the world (The Royal Forums, 2013). From the twelfth century to the mid-1800s, powerful regional warlords often rose to the position of Shogun (将軍), a hereditary military commander that sometimes wielded more power than the Emperor.

If the Opium Wars were the cataclysm that dragged China into the Western orbit, in Japan's case the national narrative acknowledges the visit of Matthew Perry of the US Navy. On 8 July 1853, Commodore Perry, commanding a squadron of two, heavily armed steamers and two sailing vessels, sailed into Tokyo Bay. On behalf of the US government, he requested Japan to enter into trade with the US and demanded a treaty permitting trade and the opening of Japanese ports to US merchant ships. The technological and military chasm between the two sides was huge. Japan had no navy with which to defend itself, and thus intimidated by the deliberate projection of military force, the Japanese had little choice but to accede to Perry's demands: an early example of 'gunboat diplomacy'. Since 1635, the *sakoku* or 'closed country' edicts and policies had been introduced in Japan; these restricted foreign relations, especially with the West, as the Tokugawa shogunate viewed Western culture and religion as hostile to the political order (Ayusawa, 1964, p. 278). Thus, for over 200 years, inbound ships were only allowed into the country from China, Korea and the Netherlands, and international trade was only permitted at Nagasaki, via a highly circumscribed protocol (Howell, 1998, pp. 73–74); although, as one would expect, there was smuggling and illegal trade, especially with China (Hellyer, 2005, p. 11).

It is the nature of the response to the stark realisation, almost at the same time, of the superior technical prowess of the West that may stand out as the defining difference between Japan and China, and which paved the way to their respective, late-nineteenth-century rise and fall. Beijing procrastinated, resisted social and political reform, and sought to maintain its undisputed position in the face of incontrovertible technological superiority and the signs of the times. The case of Empress Dowager Cixi in 1888 reportedly embezzling millions of taels of silver meant for upgrading the Imperial Navy in order to build the garden in one of her palaces is telling (Summer Palace, 2015). In stark contrast, and with the advantage of hindsight, Japan sought to protect itself from the subjugation and humiliation that happened in and to China (Refsing, 2003, p. 48). Japan witnessed the end of the last (Tokugawa) shogunate in 1868 via the Meiji Restoration (of imperial rule), which led to enormous changes in Japan's political and social structure, including conscription and land reform. Within 50 years, Japan had been transformed from a recluse state and a feudal society to a modernised nation with a market driven economy. 'Meiji' means 'enlightened rule': and this consisted in the delicate balance between modernity and traditional values. Trained in Western tactics and weapons, the Japanese army's first successful test was domestic, quashing the Satsuma rebellion of 1877, a last-ditch attempt by disgruntled samurai to stop the erosion of their powers. By 1895, Japan had secured victory against China and by 1906 against Russia. The Meiji oligarchy also set about demarcating the borders of the country: a task that had been meaningless in the previous world of tributary nations, but suddenly important in view of the Western world's expansionist mode and its interpretation of sovereignty. First in the north: Hokkaido was annexed early in the Meiji period, and its Ainu people subjected to assimilation; in the Treaty of Shimoda (1855), a Russo-Japanese border was agreed to in the Kurile Islands; and, in the Treaty of St Petersburg (1875), Japan gained the rest of the Kuriles in exchange for relinquishing Sakhalin Island to Russia. Then in the south: in 1879, the Ryukyu Kingdom was officially abolished and Japan's Okinawa Prefecture was created, much to China's frustration and prolonged protest. In 1898, Minamitori-shima, another *terra nullius*, was officially annexed. Japan expanded its territory even further following successful military campaigns. First, with the Treaty of Shimonoseki in 1895 (at China's expense); having been declared as *terra nullius*, the Senkakus were also incorporated into Japan's territory at this same time. But, in the Treaty of Portsmouth in 1905, after US mediation, and after having squarely defeated Russia, Japan was left with few territorial gains and no financial reparations. A strong nationalist and anti-US sentiment took root in the country, leading eventually to the attack on Pearl Harbor and Japan's entry into WWII in 1941.

At the start of the twentieth century, Imperial Japan had become the only non-Western world power and a major force in east and south-east Asia as a result of rapid industrialisation and economic development. Meanwhile, both Russia and China were embroiled in debilitating civil wars. Japan flexed its muscle in the resulting power vacuum: Korea was annexed in 1910; Manchuria was seized in 1931; China was invaded in 1937. Nanking, then the capital of the KMT forces in China, was captured in that year; in what is known as the Nanking Massacre, Japanese troops allegedly massacred as many as 300,000 people, including civilians, although actual numbers are uncertain, and the details of what actually happened are even more unclear (Yoshida, 2006). Then, with Japan entering WWII, Hong Kong, British Malaya and the Philippines were attacked immediately after Pearl Harbor. Singapore fell in early 1942; much of modern-day Indonesia soon followed. Japan also captured many islands in the Pacific, including Guadalcanal (modern day Solomon Islands) and an invasion of Australia was threatened. At the height of its power in 1942, the Japanese Empire controlled a land area covering 7,400,000km^2 (2,857,000 square miles), spanning Asia and Oceania, making it one of the largest maritime empires known in history.

By mid-1945, however, a thorough reversal had occurred. Soviet troops were preparing to attack Japanese forces in Manchuria and Korea. The Imperial Japanese Navy was no longer capable of conducting any major operations and an Allied invasion of the Japanese motherland itself was imminent. Two atomic bombs were dropped on the cities of Hiroshima (6 August) and Nagasaki (9 August): some 130,000 people were killed in what is still the only use of nuclear weapons in warfare. The Soviet Union declared war on Japan on 8 August, and swiftly occupied the rest of the Kurile islands. Emperor Hirohito announced the surrender of Japan to the Allies on 15 August. The Allied occupation of Japan began that same month: by the end of 1945, some 350,000 US personnel were stationed throughout the country. This marked the first time in Japan's history that the nation had been occupied by a foreign power. Its 'empire' was totally dismembered: Sakhalin and the Kurile Islands to the Soviet Union; Taiwan and Penghu to China; the Andaman and Nicobar Islands, Hong Kong and Solomon Islands to the United Kingdom; and South Korea, Okinawa, the Amami Islands, the Ogasawara Islands and Japanese possessions in Micronesia to the US (as occupying force). Japan's post-war constitution, adopted under Allied oversight, included a 'Peace Clause' (Article 9), which renounced war and banned Japan from maintaining any armed forces (other than for the sole and express purpose of self-defence).

And yet, what had originally started off as a plan to deliberately knock out Japan as an economic and industrial power was quickly revised: the rapid rise of Communism in mainland China, Russia and Vietnam, the Korean War (1950–53) and thus the onset of the Cold War, led to a radical change of plans. Japan became *the* bulwark of Western democracy in an unsettled East Asia, and has remained so. The US–Japan alliance has endured as a key plank of 'security cooperation' in the region, progressively revised to take note of domestic and regional changes. The latest (2015) update, following an Obama–Abe summit, takes into consideration Japan's increasing disposition towards military action (Hornung, 2015).

Japan thus arrives at the twenty-first century as a nation that has powerfully embraced Westernization. The country had risen to dizzy heights within a few decades, exploiting the weaknesses of regional neighbours to expand into an empire by 1942 – committing some atrocities along the way – and then quickly cut down to size within a brief three years. It aspires to recover some of its greatness in the name of *ryodo*, or inherent territory, a label applied to the Diaoyu/Senkakus, but also to Dokdo/Takeshima (controlled by Korea) and the South Kuriles (controlled by Russia) (Dudden, 2015). The rituals performed at the Yasukuni shrine in Tokyo, a central custodian of memory which honours the 'spirits' of Japanese war dead – but which allegedly whitewashes Japan's wartime brutality – help to reconstruct in the present the glories, but not the outrages, of Japan's imperial past, explaining their controversial nature (Shibuichi, 2005). The US occupation force in mainland Japan did not leave before 1952; and it did not leave Okinawa before 1972. However, the US military presence lingers on: there are still 23 US bases in Japan; and Kadena Air Force Base in Okinawa is the largest US Air Force base in the Pacific. Okinawa makes up only 0.6 per cent of Japan's land area; yet, almost two-thirds of US bases in Japan are in Okinawa; as are three-fifths of some 49,000 US troops still stationed in that country (Lutz and Enloe, 2009; *The Economist*, 2015d).

Wild card: Ryukyu/Okinawa

We are obliged to give special treatment to Okinawa in this book for a variety of reasons. It is the only island prefecture of Japan: all remaining 46 are on the 'mainland' of Hokkaido, Honshu, Shikoku and Kyushu. Shuri Castle, Okinawa, was the former seat of the Ryukyu Kingdom, a long-term tributary state of China, prior to Japanese annexation. A strong sense of Okinawan identity remains. Okinawa was the site of the most vicious battle of WWII, fought on Japanese soil and during which more than one quarter of the local population was killed. It is still swarming with US military facilities. It is indeed part of the irony of the peace settlement of post-war Japan that the country was constitutionally pacified by the US while Okinawa was heavily militarised by the same US (MacCormack and Norimatsu, 2012). Ryukyuans continue to protest the foreign military presence on the main island (also called Okinawa). Finally, it is that part of Japan closest to the disputed Diaoyu/Senkaku chain: specifically, Yonaguni (Japan's westernmost inhabited island), Iriomote and Ishigaki (and which are collectively known as the Yaeyama islands). For Japan, the Senkakus are administered as part of Ishigaki city in the Okinawa Prefecture. Various Okinawans fished in Senkaku waters and collected albatross feathers from the islands in the period 1895–1940.

The Ryukyu kingdom, a tributary state of China, secured its peak around the mid-fifteenth century. The Ryukyus thrived by developing strong trade relations with numerous other countries. This role as 'bridge' to various regional kingdoms rested on a special Ryukyu–China tributary relationship. In 1609, the island was invaded by a force from Kyushu led by the Lord of Satsuma and obliged to accept its suzerainty (Sakai, 1964). For over two centuries, the kings of Ryukyu played a risky diplomacy, paying tribute to the Japanese shogun as well as to the Chinese emperor. This double allegiance ended when Japan ordered the Ryukyus to terminate tribute relations with China in 1874. Ryukyu was annexed by and into Japan in 1879, establishing it since as the Okinawa prefecture, and abolishing its 550-year old monarchy (Yonetani, 2000). A period of Japanisation ensued: standard Japanese became the medium of instruction in schools, and the use of local Okinawan languages was discouraged; this bred widespread resentment amongst the locals (Nakasone, 2002). Then came WWII and the Battle for Okinawa in which 120,000 combatants died and Shuri Castle was destroyed. Thousands of local citizens were forced by the Japanese to commit suicide rather than surrender. At the end of the War, Okinawa was run by a US military government: this arrangement persisted even after the end of the occupation of Japan as a whole in 1952. During that time, the US dollar was the official currency, and cars drove on the right side of the road (unlike in the rest of Japan). It was largely through the US troops stationed on Okinawa, that the martial art of karate was exported to the US (Kerr, 2000). The US returned the islands to Japan on 15 May 1972, but retained its rights to bases there: the US military still controls about one-fifth of land area on Okinawa island, with some 30,000 US service personnel on active duty at any time. The US presence provides rent payments, creates job spin-offs to the locals, and supports local tourism; but there is a strong anti-US base feeling in Okinawa. The alleged robbery, choking and raping of a woman in October 2012 by US sailors led to public outrage and recalls the rape of a 12-year-old girl by three US military personnel in September 1995 which 'remains a high-water mark for anti-military sentiment' on the island (Tritten and Sumida, 2012). The scheduled deployment of 12 American MV-22 Osprey vertical take-off and landing aircraft to the Futenma base in September 2012 was the subject of large demonstrations because of safety concerns (Evans, 2015). The cases of rape, and conspiracy to rape, involving US servicemen (who are exempt from local jurisdiction), usage of banned chemicals, noise pollution and accidents involving US planes crashing into local buildings all stack up to foment antipathy. Futenma is one particularly contentious air base: it is now practically surrounded by the sprawling Ginowan City. Demonstrations on land and at sea continue in the face of plans for extending Camp Schuab, a US military base. Okinawa's governor, Takeshi Onega, elected in 2014, has pledged to stop the project (*The Economist*, 2015a). The tension has consumed not just local but even national politics. Democratic Party of Japan (DPJ) leader Yukio Hatoyama pledged to scale down the US bases on Okinawa as part of 'a more equal alliance' with the US in an electoral pledge; but, once elected as Prime Minister in September 2009, he reneged on this commitment and resigned after a few months in office (Buerk, 2010). Mass public demonstrations against the presence, even relocation and expansion, of the US bases continue: one recent demonstration, in June 2016, was triggered by the death of a 20-year-old female resident and linked to a US Marine veteran, and is claimed to have drawn 65,000 protesters (Soble, 2016). Nevertheless, it is unlikely that the Abe government would want to see US military strength in Japan diminished, in any way (Fackler, 2015a).

During WWII, a US professor with a fair knowledge of Japan had noted that 'the people of the Ryukyu Islands have a double allegiance', referring to both Japan and China (Shelton, 1945, n.p.). In the context of the Diaoyu/Senkaku affair, this situation presents both problems as well as opportunities, and the Okinawa card cannot be discounted in current China–Japan relations. First, the Okinawans are historically those Japanese citizens that have had the longest experience of living peacefully with a hegemonic China, to which they paid regular tribute. Trading links with neighbouring Taiwan were also especially extensive and strong (Lin, 2014). Second, there is an undercurrent of resentment expressed by Okinawans in regard to Tokyo: some regard the 1874 annexation of Ryukyu to Japan as a lingering colonial act that destroyed their monarchy, suspended their special relationship with Beijing, and instead thrust their island into a long period of marginalisation, a terrible world-war experience fought on Okinawan soil and the lingering obnoxious presence and deployment of foreign forces. The strong anti-US bases movement enjoys a wide appeal. A secessionist political party, Kariyushi Club, is less popular, but nevertheless is active in promoting the idea of an independent Okinawa. Third, if there are any practical benefits to be enjoyed from resolving the Diaoyu/Senkaku spat, or to be tolerated by any patch-work agreements, the immediate beneficiaries (or victims) would most likely be the Ryukyuans (and the Taiwanese) in the immediate neighbourhood of the contested islands (e.g. Asahi Shimbun, 2013). For example, in 2012, when strains in the dispute were at an all-time high, some 3,000 Chinese tourists cancelled their trips to Okinawa prefecture due to the tensions in the territorial dispute (Ryukyu Shimpo, 2012).

But, then, there are hawks apart from doves: there is a very different response to what is perceived as clear evidence of Chinese resolve and demonstration of stronger Chinese muscle power in the region. Various Ryukyuans support the deployment of additional troops and materiel to counteract the Chinese military build-up (Tiezzi, 2014). Yoshitaka Nakayama, mayor of Ishigaki City, insists that the Yaeyama islands must be protected from the Chinese threat (Warnock, 2012). Ironically enough, this looming Chinese menace, so physically close to Okinawa, has also obliged US strategists to consider their bases in Okinawa as lying perhaps too close to China for comfort: at least 5,000 US marines and 1,300 dependants will move to Guam from Okinawa Prefecture (Fuentes, 2015; *Japan Times*, 2014b).

Nakayama was re-elected mayor of Ishigaki City in 2014; his repeated demands for more central government involvement in the Diaoyu/Senkaku Islands affair have been met with calls for restraint from Tokyo. It may have been desperation that drove the 1,500 Yonaguni residents to vote (60 per cent in favour; turnout of 86 per cent) in a February 2015 referendum to support the stationing of a Japan Self-Defence Force base on the island, after years of neglect from Tokyo and its unwillingness to cooperate with the residents' plan to turn their island into an international hub (Shishui, 2015).

Conclusion: unresolved historical legacies, strong regional rivalries, booming economic relations

China is Japan's largest trading partner, and Japan is the second biggest investor in China (after the US). And yet, there is an enduring mistrust and a deteriorated perception of the other amongst the two general publics. China's rise to regional power status, Japan's lethargic performance, domestic politics and leadership personality traits have all conspired to sour relations between Tokyo and Beijing (Xiaoming, 2015). But a very heavy historical baggage also impacts on the relational behaviour of China and Japan (Qiu, 2006). Harped upon in textbooks and speeches, national humiliation – epitomised by the Massacre of Nanking – will not be forgotten easily in and by China (Wang, 2014), an aggrieved nation that has endured a century of shame and various indignities at the hands of the West and Japan (Gries, 2004; Shambaugh, 2013). Meanwhile, across the East China Sea, national humiliation is also front and centre in much political discourse. Referred to also in textbooks and speeches, and embodied in an emasculating constitution hoisted by occupying foreign powers, which prevents the country from having its own army, such shame wields much political capital in Japan (Estevez-Abe, 2014; Murphy, 2015).

Significant economic interdependence and a shared mainstream, public concern with peace and prosperity in the region characterise contemporary China–Japan relations. Yet, lingering memories of war and its aftermath have arguably been the most conspicuously contentious issue in Sino-Japanese relations for at least seven decades (Smith, 2015). The five islets and three rocks of the Diaoyu/Senkaku find themselves ensnared in the traps of recent history, where recrimination trumps reconciliation, and whose haunting recollections, it seems, neither side wishes to dismiss, forget or grow out of.

The next chapter takes a much closer look at the Diaoyu/Senkaku dispute itself, including the sequence of episodes that brought the two countries to the brink of conflict in 2012.

4 The Diaoyu/Senkaku islands saga

> Yes, the islands are small, but from a Chinese perspective, one inch of territory is big enough for China to fight for.
>
> (Cheng Xiaohe, Professor, Renmin University, China, quoted in Hatton, 2014, n.p.)

> Tokyo will defend the Senkaku Islands. It may upset other countries, but can anyone complain about Japanese people making this acquisition in order to protect Japanese territory?
>
> (Tokyo Governor Shintato Ishihara, quoted in Nakashima, 2012, n.p.)

Introduction

The Diaoyu/Senkaku islands affair has long been mired in high levels of nationalist rhetoric in both Japan and China (Chung, 2004; Downs and Saunders, 2000; Shaw, 1999). Despite efforts to avoid confrontation over this issue in the past, a series of events have led to the very tense situation that prevailed in late 2012 (see below). The vast natural resource deposits of oil and gas thought to be in the area, not to mention fisheries, may lend some urgency to bringing the matter to some resolution, while also making climb-downs by either party unlikely. Also important is to acknowledge that both China and Japan are embroiled in other, potentially more significant, island disputes in the region. China is notably at loggerheads over the South China Sea with no less than five countries (Brunei, Indonesia, Malaysia, Philippines and Vietnam); but with many non-claimants wanting the South China Sea to remain considered as international waters, and foremost amongst these the United States which continues to conduct 'freedom of navigation' operations therein. Meanwhile, Japan has its own issues over Dokdo/Takeshima with Korea and over the Southern Kuriles with Russia. Any moves and claims implicating the Diaoyu/Senkaku chain may have other and bigger prizes in mind.

Meanwhile, conventional avenues for dispute resolution exist. Since 1982, UNCLOS offers detailed technical parameters for determining the baselines from which maritime sovereignty claims ensue. The ICJ is the typical forum for the adjudication of sovereignty disputes. However, the catch here is that the Japanese government maintains that there is no territorial dispute over the Senkaku Islands, and therefore nothing to take to the ICJ. Moreover, since the 1990s, Beijing's basic policy regarding maritime territorial disputes has been to seek a resolution through direct negotiations with the countries involved, ignoring or striking out against international rulings or the involvement of a third country/party. It has opted to adopt a similar disregard with respect to a case lodged against it by the Philippines with the International Court of Arbitration in The Hague: Manila has argued that the U-shaped, nine-dashed line used by Beijing to define its claim in, and to, the South China Sea – and with a tenth dash added east of Taiwan island in 2013 – is illegal (*The Economist*, 2015f). Beijing has maintained that it would neither accept nor participate in the arbitration, having stated that the tribunal constituted at the unilateral request of the Philippines has no jurisdiction (Chinese Society of International Law, 2016). The decision on the case was announced in July 2016 (more about which below).

Former Japanese Foreign Minister Koichiro Genba appeared to signal that Japan might accept the jurisdiction of the ICJ, to which Japan has acceded generally, whereas China has long refused to accept ICJ dispute resolution (Genba, 2012). Such a route would offer some advantages to both sides, including a cooling-off period and, during the long years when the matter rests before the court, a context in which a settlement would be encouraged (Cohen, 2013). Moreover, any future willingness by the parties to agree that the Diaoyu/Senkaku islands, as 'rocks' under UNCLOS Article 121(3), are entitled only to a 22.2 km (12 nautical mile) territorial sea and not to any resource zones – such as a continental shelf or an EEZ – would go a long way to reducing the importance and significance of an eventual adjudication, irrespective of the outcome. Such a 'scaling down' position has already been taken by both the PRC and ROC.

Formal claims

The Japanese formal claim to the islands dates to the turn of the twentieth century and is based on formal occupation after Tokyo's determination that the islands were then *terra nullius*: a legal concept that refers to a territory not ruled by a recognised state, which could be the consequence of no previous discovery and claim, or of the abandonment of a previous claim (Austin, 1998, p. 113). Japan claims to have acquired title following a series of surveys between 1885 and 1895, during which the Japanese deemed there were no signs of Chinese control over the islands. Their formal incorporation was effected by a cabinet decision issued on 14 January 1895. As it so happened, this incorporation took place during the final stages of the First Sino-Japanese War, with Japanese victory in sight.

Even before 1895, there had been Japanese activity on the islands. In 1884, the Ministry of the Interior in Tokyo had been informed that there were large numbers of short-tailed albatrosses (*Phoebastria albatrus*) nesting on the Senkakus, and that these birds could be harvested for meat and feathers. In 1891, men from Kumamoto and Okinawa Prefectures started arriving on the islands, killing the albatrosses for their feathers. In 1896, Koga Tatsushiro (1856–1918), a wealthy entrepreneur, financed the indiscriminate hunting of short-tailed albatrosses nesting on Uotsurishima and Kubashima, in the Senkaku chain, for export. Some 150,000 to 160,000 albatrosses were thus dispatched. Already by 1903, the albatross-breeding colonies had been wiped out. With the supply of albatrosses exhausted, Koga shifted operations to the catching and processing of bonito, a fish commonly used in soup stocks. By 1937, the marine ecology of the islands had been stripped bare of those fish; at that point, the islands were abandoned. There has been no enduring human habitation on any of the islands since (Tyner, 2011). In 1939, not even a single short-tailed albatross was observed on the Senkakus; the same result was reported in investigations in 1950, 1952 and 1963 (Japan Reference, 2012). Some breeding pairs of albatross were sighted again on the islands only in 1971 (Ikehara and Shimojana, 1971, p. 85).

The PRC position is that these islands are historical Chinese territory, dating back at least to the Ming Dynasty (1368–1644), when they were instrumental in the country's historical tributary relationship with the ancient Ryukyu kingdom in present-day Okinawa (State Council of the PRC, 2012). Their claim is essentially based on discovery plus two acts of effective occupation: the islands' use in the sixteenth century as a navigational point in the Chinese coastal defence system, and a one-off grant of ownership to a private party by Chinese imperial edict in 1893 (Austin, 1998, p. 164). China dismisses the Japanese claim that the islands were *terra nullius*, citing Japanese contemporary references to Chinese media reports of objections to Japanese surveys beginning in 1885 prior to perfecting the Japanese claimed annexation in 1895 (Shaw, 1992). The Chinese may agree that the Japanese acquired full sovereignty of the islands in 1895, but only as part of the cession of Taiwan in the Treaty of Shimonoseki, which would have given rise to the Japanese obligation to return the islands to China after World War Two along with the return of Taiwan, under the terms of the Cairo Declaration of 1943, the Japanese Instrument of Surrender in 1945 and the San Francisco Peace Treaty of 1951. (Of course, Chinese ownership would first have to have existed for this obligation to arise). Chinese scholars have also sought to bolster the Chinese historical title claim by an irredentist argument that the then contemporary standard of discovery and claim, and not the more recent higher standard of 'effective occupation', should apply, especially in the context of uninhabited territory (Cheng, 1973-74, pp. 224–226). The weakness for this irredentist argument is that China seems to have abandoned any such claim and waited 75 years (1895–1971) before reasserting it (Suganuma, 2000); this reassertion occurred just after a 1969 UN study reported possibly large oil and gas deposits in the seas surrounding the islands: a coincidence?

Japan disputes both the claim that the Diaoyu/Senkaku islands were covered by the Treaty of Shimonoseki and the (associated) post-war obligation to restore the islands to China. Japanese officials argue that the Japanese cabinet decision annexing the islands occurred three months ahead of the Treaty of Shimonoseki and was unrelated thereto (Austin, 1998, pp. 168–170). They point out that there was *no* reference to the islands, either in the 1895 Treaty of Shimonoseki or the 1945 San Francisco Peace Treaty provisions respecting restoration of Taiwan.

(No surprise here: the islands are small, with less than 10 km^2 of land between them and held little intrinsic value, especially in the pre-UNCLOS era.) Accordingly, Japan also argues that the Diaoyu/Senkaku islands are not included in the reference from the Cairo Declaration incorporated in the San Francisco Peace Treaty requiring that 'all territories Japan has stolen . . . be restored to the Republic of China'. By Article 3 of the 1951 San Francisco Peace Treaty, Japan further agreed to the temporary 'exercise of all powers of administration, legislating and jurisdiction' by the US for the Ryukyu Islands south of 29 degrees north, under which the US took possession of Okinawa and the remaining Ryukyu Islands, as well as the Diaoyu/Senkaku islands. The US returned these islands to Japan in 1972, under the Okinawa Reversion Agreement, in spite of various 'Safeguard the Diaoyu islands' (*Baodiao*) movement demonstrations and the PRC and ROC governments registering their protest.

Crisis situations

Various crisis situations have unfolded since. In April 1978, hundreds of PRC fishing craft appeared in the waters surrounding the Diaoyu/Senkaku islands. The matter was downplayed by the PRC government, keen not to sabotage ongoing peace negotiations which led to the signing of a Treaty of Peace and Friendship with Japan in August 1978. The policy of setting aside the Diaoyu/Senkaku affair was later confirmed and reiterated in a much-quoted statement by then Vice Premier Deng Xiaoping at a media briefing event during his visit to Japan on 25 October 1978 (Tretiak, 1978):

> There exist a variety of issues between our two countries. For example, there is the issue of what is called the Diaoyu in China, and the Senkaku Islands in Japan. There is no need to raise subjects like this at a meeting like this. As I expressed to Foreign Minister Sonoda in Beijing, there is probably insufficient wisdom to resolve the issue in our generation, but with the next generation likely to be savvier than us, they will probably be able to find some resolution to the issue. It is essential to look at this issue with a broad perspective.
>
> (Ministry of Foreign Affairs of the PRC, 2014)

Then, in September 1990, reports appeared of the approval, by Japan's Maritime Safety Agency (MSA), of a new lighthouse to be erected on one of the Diaoyu/Senkaku islands as an official navigation beacon by Nihon Seiensha (Japan Youth Federation) which had already erected two lighthouses on the main island in 1978 and 1988. The Mayor of Kaohsiung, Taiwan's second city, responded by notifying the Japanese government that the Olympic Torch of the 'Taiwan Area Athlete Games' would visit the Diaoyu/Senkaku. However, MSA vessels and helicopters drove back the two Taiwanese fishing boats carrying the Olympic torch, along with reporters and TV crews. This episode was screened on Taiwan TV, leading to public uproar and anti-Japanese demonstrations. To restore some calm, Tokyo indicated in

October 1990 that it would adopt a cautious approach towards the lighthouse application and did not intend to recognise it as an official navigation beacon.

Another crisis unfolded in 1996, with Nihon Seiensha acting as the *agent provocateur*. In July, the group erected a five-metre solar powered lighthouse on one of the islands. Once again, approval was sought to have this structure noted as an official navigation beacon. The lighthouse was damaged in a typhoon, and the group returned to the island in September to make repairs. Such actions were deemed provocative by the Chinese: civil groups from Taiwan and Hong Kong sailed to the islands in protest. On 26 September, David Chan, an activist from Hong Kong, died while trying to avoid the Japanese coastguard blockade and land on the islands to plant a Chinese flag. Three days later, tens of thousands of people gathered for a candlelight vigil in Victoria Park, Hong Kong, to mourn Chan's death and protest against Japan's claims to the islands. A week later, three activists from Hong Kong and Taiwan landed on the islands and did manage to plant their flags; these were then summarily removed some days later by Okinawa authorities. Further landings by nationalists from Taiwan, but also from China and Japan, often publicised, become more frequent after 1996, raising tensions. One recent (and failed) attempt, by hot-air balloon in 2014, had Xu Shuaijun, a Chinese cook, rescued from the sea by a Japanese helicopter and handed over to a Chinese patrol ship after his hot-air balloon had suffered mechanical breakdown (Kaiman, 2014).

The next, and so far most serious, incident occurred on 7 September 2010 when a Chinese trawler, the *Minjinyu 5179*, collided – twice – with two Japanese patrol boats near the disputed islands. The Japanese authorities claim that the captain, 41-year-old Zhan Qixiong, refused to obey Japanese orders to withdraw and deliberately rammed into the Japanese vessels (McCormack and Norimatsu, 2012, p. 210). Qixiong was arrested on the spot and sent for prosecution. But the reaction from Beijing was swift and resolute: the immediate release of the Chinese national was demanded. It was also reported that China immediately blocked the export to Japan of rare-earth elements, which are essential materials in the manufacture of high-tech equipment, including military hardware, in order to exert economic pressure on the Japanese government (Bradsher, 2010; Takagi, 2016). Qixiong was duly released after 17 days in detention; he returned to Fuzhou, his home in Fujian province, on a chartered flight sent by the Chinese government on 25 September, and given a hero's welcome (Mao, 2010). Once landed, he did not miss the opportunity to make a statement: 'The Japanese capturing and arresting me is illegal. Diaoyu Islands is Chinese territory, I strongly support the government's position' (quoted in Chinasmack.com, 2010). The decision by Tokyo to release Qixiong following such threats and sanctions from Beijing was described as a 'climb-down' and 'humiliating retreat' in *The New York Times* (Fackler and Johnson, 2010).

Teetering on the brink

Even when a further escalation of the stand-off over the islands seemed unimaginable, this is exactly what happened. In September 2012, Japan's government announced that it had purchased those three islands within the group that still remained in private hands (Takenaka, 2012). This attempt was meant, Tokyo claimed, to diffuse territorial tensions and prevent even further provocations: the islands risked falling into the hands of an ultranationalist, Shintaro Ishihara, then governor of Tokyo, who had earlier indicated that he was close to buying them from their private owners (BBC News, 2012a; Smith, 2014). The decision misfired badly: Beijing protested strongly about this act that once again implicitly recognised Japanese sovereignty. Within a few days, Chinese surveillance ships sailed into waters around the Diaoyu/Senkaku islands, asserting China's territorial claims. Meanwhile, huge anti-Japanese protests broke out in cities across China. The Japanese embassy in Beijing was besieged by thousands of protesters throwing rocks, eggs and bottles. In some cities, peaceful protests turned violent as protesters clashed with the police, attacked Japanese-made cars and smashed up Japanese restaurants. Some major Japanese firms, such as Toyota and Honda, temporarily shut factories and offices across China. Chinese coastguard vessels have since maintained a regular presence in the waters around the Diaoyu/Senkaku islands, where a cat-and-mouse game ensues with their Japanese counterparts. Meanwhile, airspace has also been embroiled in the stand-off. In August 2013, a Chinese Air Force fighter jet coasted along the airspace of the Diaoyu/Senkaku islands, sparking the immediate reaction of the Japanese Self-Defence Forces.

By November 2013, China had set up an 'air-defence identification zone' (ADIZ) over an area of the East China Sea, and covering the disputed islands. Any aircraft entering the zone must now report flight paths, maintain two-way communication or else face 'emergency defensive measures' (Xinhua, 2013). Yet, Tokyo insists that it rejects the ADIZ, and no Japanese civilian flights are known to file their passage as requested. The US has toned down its initial opposition to the ADIZ, but US military planes do not obey Chinese demands of informing and identifying their routes prior to entry into the ADIZ set up recently by China, and which overlaps with Japan's own ADIZ and includes the disputed islands. Indeed, at least two US strategic bombers have intruded into this airspace on 26 November 2013, a good number of the intrusions continue to happen, and quite a few of them are met with emergency retaliatory interceptions by scrambling Chinese fighter jets (Zhu, 2014).

The political climate has improved slightly since. In November 2014, the leaders of the two countries declared that they were trying to roll back a prolonged stand-off that has inflamed nationalist sentiments, damaged economic ties and at times appeared to bring them close to military conflict (Perlez, 2014). But the manoeuvres continue: in cyberspace, on the sea and in the air. Both China and Japan have taken to the Internet to promote their territorial claims, rolling out detailed websites that document their official positions (Tiezzi, 2015). The sale of 17 V-22 Osprey military transport aircraft and associated equipment by the US to Japan was approved in 2015, adding to Japan's F-15 fighter jet fleet already stationed in Okinawa (Gady, 2015). Meanwhile, China is deploying its J-7 interceptors, J-10 fighter jets and the multi-purpose Shaanxi Y-8 aircraft to patrol the skies over the East China Sea (Johnson, 2013). In the context of the current impasse and *de facto* stalemate, pushing the wrong button on the control panel of a coastguard vessel or on a military aircraft may have very serious consequences.

Conclusion

In any case, and as long as no such human error is made, a set of carefully calibrated responses and counter-responses unfolds like a ritual in the East China Sea. Beijing and Tokyo now appear to be following an agreed-on protocol of patterned patrols around the disputed islands. These actions have not resolved the dispute; but they have at least '. . . kept the situation from boiling over and made it more tractable' (Christensen, 2015b, p. 32). However, while Chinese patrol ships have occasionally sailed close to or entered what Tokyo considers its territorial waters around the Diaoyu/Senkaku island chain, a Chinese naval vessel made its way into those waters for the first time in June 2016, leading the Japanese government to summon the Chinese ambassador in Tokyo and protest formally about the incident (*Guardian*, 2016).

This chapter has thus scoped the details of the Diaoyu/Senkaku dispute and the respective competing claims and counter claims of Tokyo and Beijing. So far, there is nothing really original: the collation of facts and most of the arguments proposed herein have been repeated and reproduced *ad nauseam* on websites and in publications. What this text now goes on to do is to suggest a comparative island framework as a new, alternative and compelling conceptual and analytic perspective. It is precisely by drawing other islands into the mix, and by observing critically the solution protocols that have emerged from their particular circumstances, that some creative considerations to the Diaoyu/Senkaku case can be proposed.

5 Worth nothing, yet worth everything

Why countries (may) go to war over small islands

Introduction

The front cover of the *Economist* magazine showed a clutch of small rocks and uninhabited islets, with a collective land area of less than 10 km^2. In complete disproportion to their insignificant size, the caption suggested that China and Japan may 'really go to war over these' (*The Economist*, 2012a).

This set of islands is located in the East China Sea, between China and Japan. They are uninhabited. They are currently notionally administered by Japan as part of Ishigaki City, Okinawa, and which calls the islands *Senkaku* (尖閣諸島). This name is a direct translation from the English name Pinnacle islands, which was first used by Captain James Colnett of the *Argonaut*, who charted them during his 1789–1791 voyage. But the islands are also claimed by China/PRC by whom they are called *Diaoyu dao* (钓鱼岛). This name is actually the name of the largest island in the group, and which translates as a fishing platform. The islands are also claimed by Taiwan/ROC, by whom they are called *Diaoyutai* (釣魚台列) respectively, and which carries the same meaning.

The Diaoyu/Senkaku islands actually consist of five islets and three rocky outcrops. To the Chinese, these are respectively: Diaoyu Dao (the largest), Chiwei Yu, Huangwei Yu, Bei Xiaodao and Nan Xiaodao, along with Bei Yu, Nan Yu and Fei Yu. To the Japanese, these are: Uotsuri-shima (the largest), Taishō-tō, Kuba-shima, Kita-Kojima and Minami-Kojima, along with the rocky outcrops of Okinokitaiwa, Okinominamiiwa, and Tobise.

No one currently uses the islands for anything. But, 'if World War III takes place anytime soon, this is where it will start; implausible as that may sound' (Posner, 2014, n.p.).

Why indeed should a few rocks stir such serious passions and lead to such dangerous confrontations, which could also spark regional, even global, conflict? How is it possible that regional powers with stable governments, who are also neighbours and important trading partners, squabble over such seemingly minor issues? And why are island disputes especially difficult to resolve?

This chapter looks beyond the immediacy and topicality of the Diaoyu/Senkaku affair. It first espouses how various small islands throughout history have ended up in a similar predicament, often being sucked into – and often ending up being devastated by – regional conflicts, with the contending powers not necessarily interested in the islands per se, but rather in what they portend, strategically and psychologically. A comparison and contrast is made between the Diaoyu/Senkaku case and the situation of another contested group of islands: the Falklands/Malvinas.

Small islands and a tense presence in history

Since the territorialisation of states following the Treaties of Westphalia in 1648, it should come as no surprise that many states experience, or *claim* to experience, threats to their cohesion and political integrity. This challenge is not just a possible outcome of the misalignment between a place-based state and the nation as a symbolic unitary public (Anderson, 2006; Guibernau, 2013). Large states with sprawling geographies and heavily populated communities, possibly also split along multiple ethnic and racial lines, seek to balance these centrifugal forces with federative arrangements, granting or condoning some level of devolution and/or self-determination to stave off demands for secession, or to bring together previously self-governing territories (Ghai, 2000): Brazil, Canada, India, Nigeria, Russia and the United States are good examples. In other, also largely continental, states, similar expressions of local identity exist; but, this time, these are organised and coalesce around distinct *island* affiliations and sympathies, often bred out of specific historical and/or (post)colonial experiences that also nurture a 'love–hate' relationship between the island and its respective mainland: Åland (with respect to Finland); Sicily and Sardinia (Italy); Corsica (France); Jeju (Korea); Hawai'i and Puerto Rico (USA); Newfoundland (Canada); as well as Macau and (part-island) Hong Kong (China) are suitable examples of island pseudo-ethnic nationalism (e.g. Baldacchino and Hepburn, 2012; Vézina, 2014). In four countries – Indonesia, Japan, the Philippines and the United Kingdom, which are the world's only large and heavily populated archipelagic states – these two forces are combined, to varying degrees. Tensions in Irian Jaya, Indonesia (May, Matbob and Papoutsaki, 2013); the independentist Kariyushi Club ((かりゆしクラブ, Kariyushi Kurabu) in Okinawa, Japan (Ishiyama, 2015); the Muslim separatist movement in Mindanao, Philippines (Podder, 2014); and the 2014 Scottish independence referendum in the UK (Rosie, 2014) are all keen reminders of this centrifugal tension in subnational island jurisdictions, expressed in different ways, some peacefully, some more violently.

Nor is this edgy existence for small islands a recent historical phenomenon. In the island-rich Caribbean, many of what today are small island states have changed hands in the context of regional power politics between the sixteenth and nineteenth centuries: for example, St Lucia changed hands between the French and the British no less than 14 times. Even earlier, the islands of the Mediterranean were already testimony to considerable military operations, often involving the military powers of Christendom versus Islam, but also the crafty commercial interests of city states like Genoa, Pisa and (most successfully) that island city state, Venice (Abulafia, 2011).

In many such and other cases, small islands are mere pawns in a larger geostrategic conflict. They are *not* seen as important acquisitions for their own sake. Rather, they are perceived as 'bargaining chips', assets that help one party to improve its negotiating position vis-à-vis another; that secure its logistic, resourcing and/or refuelling challenges; or that get its military *materiel* much closer to its enemy's heartland. Islands also become small prizes that however can score important psychological victories; and they are tactical battlegrounds that force the enemy into pouring kit, soldiers and other resources that are then 'locked' and may not thus be used in other, more significant theatres of war. Small islands can serve as *tactical feints* (such as to lure or distract observers away from more important targets or issues); they can function as *proxies* (dummies planted by third parties to forestall cooperation among two regional players); they can work as *pretexts* (sites for a 'dress rehearsal' in anticipation of bolder and broader territorial ambitions); and also as *bait* (meant to test the resolve of other parties towards action). Just to give an example of the lingering and sheer value of islands in current military strategy, as of 2002, in 29 out of the 59 countries where the US had an extra-territorial military presence, such bases were located on islands, and sometimes took up entire islands (Foster, 2006).

Consider the following historical episodes as poignant examples of the significance of small islands, not for what resources the islands contain or command, but for their relevance to 'big picture' political, nationalist and/or geostrategic issues.

The 'Great Siege' of Malta (1565)

While significant to the history of the Maltese islands (land area: 316 km^2), this episode was but one minor act in the long-term clash between Christendom and Islam in the Mediterranean basin (Cassola, 1995). Having been evicted from Rhodes by the Turks after a siege in 1520, the Knights Hospitaller of St John, a medical-military Christian outfit for aristocratic Europeans, took up residence on the Maltese islands in 1530, gifted to them by Emperor Charles V of Spain. Their regular harassment of Turkish shipping from their new base in Malta eventually led Ottoman Emperor Suleiman to send out an expeditionary force to destroy the Knights there. In spite of an Ottoman army of some 30,000 strong (including 6,000 crack Janissaries) facing 500 Knights and some 6,000 soldiers, the Turks had to abandon a siege after three months. Almost a third of the islanders perished. The failure of the siege did not reverse the increasing dominance of Ottoman naval power in the Mediterranean at that time; it took place at a time of Christian defeats, including the Battle of Djerba (1560) and the loss of Crete (1571) (Bradford, 1961).

Midway, Peleliu, Iwo Jima, Okinawa (1942–1945)

These were the island sites of some of the most significant conflicts between Japanese and US forces in the Pacific theatre of World War Two (1941–1945) (Costello, 1982).

At the Battle of Midway (3–6 June 1942), the US sunk four of Japan's eight aircraft carriers (for the loss of one of its own) and effectively ended the threat of further Japanese invasion in the Pacific. The engagement started with Japanese plans to capture the atoll (land area: 6.2 km^2). Tokyo had plans to transform Midway into a forward outpost from which Japan could eliminate any future US menace in the Central Pacific.

At Peleliu (September–November 1944), US troops successfully captured an airstrip on the small coral island (land area: 13km^2), part of modern-day Palau (Moran and Rottman, 2002). The airstrip was deemed strategically critical for an eventual invasion of the Japanese mainland.

US military forces invaded Iwo Jima, a small volcanic island (land area: 21 km^2), in February 1945, as part of their island-hopping strategy to defeat Japan. They secured its three airfields for launching further operations, but only after a vicious 35-day battle.

Finally, the US planned to use Okinawa, a relatively large island (land area: 1,207 km^2) just 550 km away from mainland Japan, as yet another base for air operations on the planned invasion of the Japanese mainland. Starting in April 1945, a long and nasty land battle ensued, with over 200,000 soldiers and islanders killed. Some fighting continued even after Japan's unconditional surrender in August 1945.

Kinmen and the Battle of Guningtou (1949)

Kinmen (which means golden gate; also known in the local dialect as Quemoy) is governed by Taiwan. It is an island chain with a resident population of around 120,000 (land area: 150km^2). Kinmen's very particular location assumed strategic significance after 1949: some of its islands are located less than 1 km from the territory of the PRC – it is located about 2 km (1.2 miles) east of the mainland Chinese city of Xiamen – while lying more than 200 km from its 'mainland', Taiwan. The ROC initially saw Kinmen as the launching pad for the hopeful invasion and resumption of its control over mainland China. But the PRC regime in Beijing targeted it as the obvious stepping stone for an eventual full-scale invasion of Taiwan. The island was thus the site of a historic battle of Guningtou on 25–27 October 1949, a botched invasion that halted the Communist advance. This event represented a powerful psychological lift in an emerging Cold War scenario (Szonyi, 2008). Kinmen remained a tense frontier site between the 'free world' and the Communist regime during the Cold War (Myers, 1996). The PRC's other amphibious operations involving islands were more successful, leading to the Communist conquest of Hainan Island, the Wanshan Islands and Zhoushan Island (all in 1950) (MacFarquhar et al., 1991, p. 820).

The Cuban 'missile crisis' (1962)

Cuba is not a small island: it is the 17th largest island in the world, and the largest in the Caribbean. It is also a sovereign state, home to a resident population of around 12 million. The missile crisis was an edgy two-week (14–28 October 1962) showdown between the US and the USSR. Moscow had agreed to Cuba's request to place nuclear missiles on the island as a deterrent to a possible US invasion. Missile launch facilities were constructed on Cuba and discovered by the US following a spy-plane mission. Washington then established a military blockade to prevent further missiles from entering Cuba and demanded that any existing missiles be dismantled. An agreement was secured after a period of tense negotiations, during which the world seemed on the brink of a nuclear war. The event loomed far beyond that Caribbean island and was a touchstone in US–USSR relations (Allison and Zelikow, 1999; Chayes, 1974).

Isla de Perejil/Leyla (2002)

Isla de Perejil (literally, Parsley Island), or Leyla in Arabic (actually, a corruption of the Spanish La Isla, which means 'the island'), is a small, uninhabited rocky islet with a land area of 15 hectares (or 37 acres), located in the Strait of Gibraltar, just 200 metres from the Moroccan coast, 8 km (5 miles) from the Spanish enclave of Ceuta and 13 km (8 miles) from mainland Spain. The islet's sovereignty is disputed by Morocco and Spain. On 11 July 2002, a group of Moroccan soldiers set up base on the islet, later replaced by cadets from the Moroccan navy, who installed a fixed base there. Spain and various other countries protested against these actions. On the morning of 18 July 2002, a Spanish commando attack force took control of the island and custody of the Moroccan naval cadets, who had not resisted the attack. Spanish troops remained on the island until Morocco, after mediation by the US, agreed to return to the *status quo ante bellum*. The islet is now once again deserted (La Parra Casado et al., 2007).

Total island population removal

Moreover, small islands suggest quick and total military operations. Once an enemy force has reached a small island, the usual outcome is eventual capitulation of the local garrison (Royle, 2001). Indeed, it was also not uncommon throughout history for *complete* island populations to be dragged into slavery. In 1544, Hayreddin Barbarossa raided the island of Ischia (part of modern-day Italy) taking 4,000 prisoners, and enslaved some 9,000 inhabitants of Lipari, that island's almost entire population. In 1551, Turgut Reis enslaved the entire population of the Maltese island of Gozo, about 6,000 persons, sending them to Ottoman Tripolitania. Even as late as 1798, the island of San Pietro, south of Sardinia, was ransacked by a surprise pirate raid and all 920 survivors, mainly women and children, were taken away and sold as slaves in Tripoli (Baldacchino, 2015; Zacks, 2005). And, in the modern age, complete island populations have been forcibly removed in the name of military strategy – such as from Diego Garcia, British Indian Ocean Territory (Vine, 2011) – or in the name of science and nuclear arms testing – such as from Bikini atoll, Marshall Islands (DeLoughrey, 2013).

Islands as dynamic arenas of political transformation and economic value

Various islands find themselves similarly, awkwardly placed: sitting uncomfortably at the cross-hairs of competing regional powers, through no fault of their own. In an atlas of 123 'geo-political hotspots', Anderson (2000) identifies 21 cases that involve islands, in whole or in part. Moreover, and possibly counter-intuitively, members of the general public who live closer to disputed islands are more likely to be pragmatic and support a compromise than those who live further away, because the former are more impacted by economic considerations. In contrast, for those living far from disputed island territories, their symbolic and psychological value is strongest. They can therefore afford to focus their attention on the purely political aspects of the dispute, which leads to a more hawkish, less conciliatory, stance (Tanaka, 2015).

A typical situation is for an island, or a group of islands, to be administered and governed by one country; but is then claimed by some other neighbouring country (or countries). Again typical of these situations is that *the whole* island, or a whole archipelago, is the subject of dispute: it is as if having only one jurisdiction on and for an island is a natural occurrence. Indeed, while there are thousands of inhabited islands in the world today, *only ten* are divided between more than one country. Three of these ten cases – Cyprus, Ireland, Timor – are included amongst Anderson's 21 cases as extant geopolitical island hotspots. Thus, except for these few exceptions, small islands have inspired 'dreams of total knowledge' and 'delusions of comprehensive totality' (Macfarlane, 2012, p. 111) – serving as convenient microcosms of the wider world – as much as misapprehensions of absolute, indivisible and total control. In fact, as much as in fiction, small islands fit the typical Robinsonade narrative and have a habit of breeding despots, who are typically males (Baldacchino, 2012b; Lawrence, 1928; Redfield, 2000, p. 12). As Dyer (2013, n.p.) has noted, an obvious solution to the Diaoyu/Senkaku affair would be to divide the disputed territory evenly. But, it seems that emotions run riot where islands are involved:

> The Japanese and the Chinese could [do it]: no face lost, and everybody makes a profit . . . So why don't they just do it? Maybe because there are islands involved . . . [A]ctual islands, sticking up out of the water, fall into the category of 'sacred national territory, handed down from our forefathers' over which large quantities of blood can and must be shed.

Growing islands

Meanwhile, ocean spaces are no longer dismissed as 'empty' but have become 'a dynamic arena of political transformation and claims-making' essential to the dynamics of late capitalism (Chalfin, 2015, p. 101). Islands have become and remain critical in maritime and territorial disputes, particularly with regard to their capacity to generate jurisdictional claims for the delimitation and expansion of maritime boundaries. International regimes like UNCLOS, of which both China (but not Taiwan) and Japan have been signatories since 1996, have 'vested otherwise worthless islands with immense economic value' (Ramos-Mrosovsky, 2009, p. 906). Suitable examples include: Heard and McDonald islands, Australia (Kriwoken and Holmes, 2007); Johnston Atoll, USA (Van Dyke et al., 2010); and Clipperton, France (Pauly, 2009). Each of these small islets has provided the legal justification to the extension of the exclusive economic zone of their respective sovereign state by many thousands of square nautical kilometres. In contrast, the island of Rockall provides the only example to date of a state (the United Kingdom) voluntarily downgrading an insular feature to a 'rock' when it acceded the UNCLOS Convention in 1997, and thus reducing the area of its claimed maritime zone (MacDonald, 2006).

Okinotorishima (沖ノ鳥島)/Douglass Reef

In Okinotorishima, Japan has its own notable example of a small islet which, however, comes with a large EEZ concession. Declared Japanese territory in 1931, this is a coral reef about 1.7 km north to south and 4.5 km east to west, some 1,700 km south of Tokyo (De Meyer, 2011). Within the circumference of the reef, and surrounded by circular concrete breakwaters to guard against waves, sit Higashi-Kojima (eastern islet) and Kita-Kojima (northern islet). At high tide, one islet has an area of 1.6 m^2 (roughly the size of a twin bed) and pokes just 7.4 cm out of the ocean; the other islet is 7.9 m^2 (the size of a small room) and rises 16 cm). But such tiny dimensions can deceive by 'being above water at high tide'. Tokyo claims that these objects meet the requirement for being called 'islands' as stipulated by UNCLOS: this lets Japan claim 'territorial waters' within a 22 km (12 nautical mile) radius of the islands, and a 370.4 km (200 nautical mile) 'EEZ' encompassing an additional 390,000 km^2 (150,000 square miles), strategically located between Taiwan and US military bases on Guam (Silverstein, 1990). (Japan's latest extension to its EEZ followed a 2012 UN commission decision: this approved the Shikoku Basin and the Southern Oki-Daito Ridge regions as 177,000 km^2 additions (Ryall, 2014)). Six coral outcroppings had been visible above the water line at Okinotorishima in the past; however, constant erosion and pounding by the waves have caused some to disappear. Fearing the remaining islets could be lost to further erosion, the central government erected protective concrete seawalls and iron breakwaters in 1987. Construction was also started in 2011 on a US$790 million harbour facility, scheduled for completion by 2016, but delayed because of a serious accident (*Japan Times*, 2014a). Additionally, efforts have been under way for several years to propagate 'star sand' coral (*foraminifera*) within the lagoon to make the atoll grow 'naturally' into an island (Wittmeyer, 2012).

China does not dispute Japanese territorial claims to Okinotori island. However, Beijing (along with Seoul) has complained about regarding the Higashi-Kojima and Kita-Kojima rocks as islands, claiming that they 'cannot sustain human habitation or economic life of their own', as stipulated in the UNCLOS island regime, and therefore cannot be used to extend Japan's continental shelf or exclusive economic zone (Ryall, 2014).

Yongshu Island (永暑岛)/Fiery Cross Reef

And yet, people in glasshouses should not throw stones: China must be wary here because it is taking its own initiatives to grow and build up other islands elsewhere. In the South China Sea, a feverish occupation of reefs and islets is under way, including ambitious land reclamation projects by China in places like Fiery Cross Reef, also known as Yongshu, in the Spratly island chain; and which, by the way, is also claimed by the Philippines and Vietnam (Lerhe, 2015; Perlez, 2015). Beijing established its first permanent base in the Spratly chain on Fiery Cross Reef in the late 1980s; and the transformation of the reef since then has been nothing short of phenomenal. Reclamation activity in the area has led to the construction of a 3,000 m airstrip and a naval facility in this location (Hardy and O'Connor, 2014; Pao-Min, 1990). Meanwhile, reclaimed sand has been added to Hughes Reef to support a lighthouse and a defence post (O'Connor and Hardy, 2015). Similar additive work is under way at Gaven, Cuarteron, Eidad and Mischief Reefs (*The Economist*, 2015b).

The regime of islands, as provided in Article 121 of UNCLOS, remains unclear, and no authoritative ruling or consensus on its interpretation has yet emerged. Depending on circumstance, including their ability to support habitation, islands may theoretically have a full, partial or no effect on the determinations of entitlement to maritime areas (Schofield and Wang, 2012). This situation has various states scrambling to strengthen their claims – enlisting sand additive and coral seeding in the process – that offshore islands are theirs, and can indeed sustain human habitation or economic life. The huge swathes of ocean that such claims may harness for island and coastal states is considerable: these spaces would be prized areas for fishing, for the exploration and extraction of mineral deposits on or under the seabed, and perhaps even for sand itself (*The Economist*, 2015c).

Nationalism and island spaces: the Falklands/Malvinas

Islands that lie 'at the edge', geographically and politically, as much as psychologically, of imagined national or imperial boundaries, are also bound to excite nationalist passions (Davis, 2015). Strident discourse from state agencies, possibly fanned and supported by populist media and local right-wing associations, can very easily craft a narrative that speaks of such islands, and any of their inhabitants, as integral to the very soul and identity of the people/nation. (Ironically, such expressions often fly in the face of actual policy and historical record, which may be best characterised by long periods of official lethargy and benign neglect.) Sentiments of territoriality can find expression in outward displays of emotion and affect, not just in public demonstrations of loyalty and patriotism, but also in just-as-manifest spectacles of antagonism to the vilified out-group: the party that is seen to be threatening the integrity of the nation. These highly charged episodes can lead to symbolic violence (such as effigy or flag burning), physical violence (such as vandalising property, beatings and lynch mobs), as well as state-sanctioned armed conflict and aggression, with the deployment of troops, warships and military aircraft. Depending on their outcome, these acts tend to either bolster or totally destroy the popularity of incumbent politicians and governments. But, they also risk getting out of control; nor are the scars of such conflicts easily healed or forgotten.

A suitable example of such a dispute concerns the Falkland Islands or simply Falklands (to the United Kingdom) or Las Malvinas (in Spanish, to Argentina). London's resolute defence of a distant people as an 'island race' and whose 'way of life is British' (Short and Carrington, 1996, p. 66) followed the Argentine 'invasion' of the islands in 1982. This episode helped UK Prime Minister Margaret Thatcher to remain in power following her electoral victory in June 1983 (Clarke et al., 1990; Dillon, 1989). The archipelago, referred to by US President Ronald Reagan as 'that little ice-cold bunch of land down there' (Charlton, 1989, p. 58), lies in the South Atlantic Ocean, some 640 km northeast of the southern tip of Argentina, but some 12,800 km from the UK. Official documents suggest that a military conflict between Argentina and the United Kingdom in the early 1980s over the Falklands/Malvinas was unlikely. 'For the British government, the Falklands were about as low as they could be on its list of priorities' (Hobsbawm, 1983, p. 14). In a move that would have been interpreted in Buenos Aires as evidence of abandonment, the ice patrol vessel *HMS Endurance*, the only British vessel in the region, was targeted for withdrawal and scrapping in 1982, and with no replacement in sight. It was the April 1982 Argentine 'invasion and occupation' of the islands that brought the very existence of the overseas territory to the attention of many people in the UK for the first time. Not only were the Falklands/Malvinas marginal to the United Kingdom's international concerns prior to the invasion; but successive UK governments had even appeared to be moving towards an accommodation with the demands of the Argentine ruling junta led by General Leopoldo Galtieri for a full transfer of sovereignty (McCourt, 2011). The history of Britain's dealings with the islanders signalled to many that the islands were just 'an unwanted problem' (Freedman and Stonehouse-Gamba, 1982, pp. 7–11).

There are many differences between the Falklands/Malvinas and the Diaoyu/Senkakus. First, and crucially, the Falklands/Malvinas have been populated for about 200 years, and with a staunchly pro-British citizenry. This creates a voluble third party to the dispute, and which is not to be taken lightly: in a March 2013 referendum organised by the Falkland Islands Government, 1,513 out of 1,517 voters indicated that they preferred that the Falklands/Malvinas remain a UK overseas territory (with a voter turnout of 91.4 per cent). Meanwhile, Japanese (mainly Okinawans) lived on the Diaoyu/Senkakus from 1895 until 1940, to harvest albatross feathers. A factory was also built to process *katsuobushi* – dried, smoked bonito fish – out of which *dashi*, a staple item in the Japanese kitchen, is made. The population of the largest island, Diaoyudao/Uotsuri, topped 200 at one point (Tadashi, 2013; *The Economist*, 2012b). But no one has lived or worked on the islands for over seven decades.

Second, the 're-invasion' and 're-occupation' of the islands by the UK has been justified in order to defend a certain *idea* of Britain in international affairs. London has not wavered in its claim that the Falkland Islanders have the right to self-determination and that, as long as the islanders want to remain British, then the UK government will support and protect the islands. There is, unlike Buenos Aires, and unlike Beijing and Tokyo in the Diaoyu/Senkaku case, no passionate pledge to defend national, sacred ground coming from London.

Third, both parties to the dispute admit that they have important policy differences over the islands; whereas Japan continues to officially claim to have no dispute with China over the Diaoyu/Senkakus.

And fourth, Argentina and the UK *did* go to war about their island tiff, leaving 649 Argentine and 258 British soldiers killed. China and Japan have come close with respect to theirs, but (so far) have stopped at the edge of the proverbial cliff.

Nevertheless, there are also some uncanny similarities. First, we are referring to two sets of islands with rival historical claims, in some cases going back centuries, and which explains the alternative naming preferences. Second, we are talking about archipelagos worth nothing much in themselves, but each with their rich fishing grounds and potential sub-sea resources in their respective EEZ. Indeed, the South Atlantic island dispute developed fresh tensions between the contending parties with the discovery of natural resources as first, rich stocks of ilex squid (*Ilex argentinus*), and then oil and gas, were found in Falklands/Malvinas waters (Ficenec, 2014; Critchlow, 2015). Third, as the dispute continues to fester, other regional powers are being roped in. For example, MERCOSUR, a four-country South American trading bloc (which includes Argentina), has agreed to close its ports to ships flying the Falkland Islands flag (BBC News, 2011). (As an interesting connection between the two cases, China has pledged support to Argentina in its claims over the Malvinas, while Argentina acknowledges China's sovereignty over the island of Taiwan.) Fourth, and perhaps most importantly, Diaoyu/Senkakus and Falklands/Malvinas are both examples where it is largely domestic political concerns that have shaped foreign policy decisions, including those entailing high risks (Lebow, 1983, p. 32). No wonder that Dyer (2013) asks whether the Diaoyu/Senkakus constitute 'another Falklands'.

The similarities extend also to the universe of creative, possible solutions. In a vein similar to that of this book, Tenembaum (2012) has advocated 'theoretical solutions' to the Falklands/Malvinas conflict. These include 'some kind of condominium or joint sovereignty' over the islands, whereby Buenos Aires could claim sovereignty (albeit a shared one), while the UK would not renounce its own legal rights and claims. A second solution sees the Falklands/Malvinas becoming fully autonomous and demilitarised within a loose federative structure. 'Various forms of joint sovereignty scenarios can be envisaged' (Tenembaum, 2012, n.p.); and UK MP George Galloway has opined that such an arrangement would be even in the UK's long-term interests (PressTV, 2013). The condominium idea, 'joint administration' and other options were actually explored during bilateral negotiations between Argentina and the UK in the mid-1970s, but were resolutely derailed by a strong and active Falkland Islander lobby in London (Donaghy, 2014, pp. 24–27; Eddy and Linklater, 1982, pp. 16–17).

Conclusion

The Falklands/Malvinas affair offers a recent historical example of how a small island chain can draw countries to war, with grave loss of life and lingering animosities thereafter. The implications of UNCLOS on maritime affairs have catapulted small islands to major economic (apart from geo-strategic) significance in the twenty-first century. Yet, even in the throes of such unfolding episodes as the Falklands/Malvinas, 'joint administration' and the condominium options were some expressions of solution protocols that have been raised, in formal or informal discussion. Aware of the riskier and more tragic alternatives, contending parties have not shirked away from exploring *at least the possibility* of win–win solutions. The next chapter introduces the notion of 'creative solutionism' and the various permutations of this notion, as exemplified in the handling, even the resolution, of numerous island disputes.

6 A genealogy of creative governance

Introduction

Forty-three of the world's 193 sovereign states and full members of the United Nations (22 per cent of the total) are exclusively island or archipelagic territories (CIA, 2015). Of these, 39 are members of the Alliance of Small Island States (AOSIS, 2015). In spite of their small size, these countries have often grabbed media attention, as in the 15th 'Conference of the Parties' (COP 15) in Copenhagen in 2009, and COP 21 in Paris in December 2015, deliberating international responses to global warming and sea-level rise (Royle, 2011, p. 250; Aosis.org, 2015). Some of these states have been small enough to lend themselves to interesting political experiments: as with the New Hebrides, the colonial name for an archipelago in the Pacific Ocean that has been independent since 1980 as Vanuatu; or Tuvalu, one of the world's smallest archipelago states, also in the Pacific. (More about both of these, and more, below.)

But the world is not made up only of sovereign units. Much less visible (albeit more common) is an additional cluster of *sub-national jurisdictions* (SNJs): many of these are islands. Over a 100 *sub-national island jurisdictions* (SNIJs) enjoy varying degrees of autonomy but fall short of political independence, so they do not enjoy (full) sovereignty (Baldacchino and Milne, 2006; Rezvani, 2014; Stuart, 2008; Watts, 2008). Some of these island cases may be little better than privileged local authorities within a sovereign state (as with Saba and Bonaire, special overseas municipalities of the Netherlands). Then, there are a variety of units – provinces as in Canada and China, regions as in Russia, states as in Australia and the US, prefectures as in Japan – that exist within federal arrangements, and typically within territories that have a large land mass and/or a large population (think of the island province of Hainan in the PRC, the province of Prince Edward Island in Canada, the state of Hawai'i in the USA, and the island prefecture of Okinawa in Japan). Next, there are 'special regions' that benefit from distinct legal provisions (like inter/national treaties), as apply in the case of Lord Howe Island (Australia), the Channel Islands and the Isle of Man (British Crown), Svalbard (Norway) and Åland (Finland). Other cases respect a distinct history, a particular culture, and/or a nuanced economic regime (as with the 'special administrative regions' of Hong Kong and Macau within the PRC). Most of the remaining examples of SNIJs would have been called 'colonies' until recently: remnants of former empires or trust territories that enjoy locally representative government, yet maintain strong ongoing relationships with Amsterdam, Canberra, Copenhagen, London, Paris, Washington, DC or Wellington (such as the Cook Islands, Falklands, Greenland, Niue and Sint Maarten/St Martin). In spite of their diversity, these SNIJs have evolved to address common challenges: nationalistic compromise, exploiting the economic interdependence ushered by globalisation, providing advantageous opportunities to continental/larger sovereign states (such as preferential access to natural resources, access to distant military base facilities and extended sovereign reach) that were, and are now no longer, available through forcible imperial control (Rezvani, 2014, 2015).

Additionally, there are smaller pieces of territories, again mainly islands, but also continental enclaves/exclaves, that may have no particular jurisdictional capacity, but that nevertheless have been bestowed with, and enjoy, some kind of special status within a sovereign state. Many of these lack the *need* for political autonomy because they are uninhabited. Nevertheless, they have been excised by their central government to perform some specific function. Their separateness, distinctiveness and more manageable smaller size, plus their indigenous peoples, idiosyncratic biota and/or habitats, renders them, for instance, obvious destinations of choice for designating aboriginal enclaves or sustainable ecotourism programmes through biosphere reserves, national parks and other diversity-rich areas (Di Castri and Balaji, 2002). Some 166 of UNESCO's World Heritage Sites are on islands; of these, at least 27 are whole islands or island chains (Baldacchino, 2010, p. 16). Fraser Island (Australia), James or Kunta Kinteh island (Gambia), Rapanui/Easter Island (Chile), Skellig Michael (Ireland) and Venice (Italy) are random examples.

By means of acts of idiosyncratic and often asymmetric governance, and adroit manoeuvres of political strategy, specific places can be transformed into something distinct from the mainstream. The distinctive physical and geographic circumscription of islands – fashioned as they often are, by nature; but also artificially constructed nowadays with current technology and engineering expertise – makes these islands especially attractive and viable locations for such projects. Thus, acts of excision, zoning, detachment, outbordering, quarantining or offshoring would faithfully describe some of these broadly similar processes or strategies. Deliberate or fortuitous, the method involves the endowment of a specific place with a particular and closely circumscribed status, and with accompanying privileges and powers, often ratified by law. Since such projects in creative governance often advise both containment and distance, a 'test site' on an island, or an island *in toto*, makes the perfect candidate.

Islands, thus, lead the way in these political adventures. The world's first free 'export processing zone' (EPZ) came into being in 1937 on Staten Island, New York, USA (Palan, 2003, p. 119). The model was then perfected with respect to Puerto Rico, one of the world's largest SNIJs, which still tests the aptitude for creative governance amongst US lawmakers (Rezvani, 2014; and about which more below). Other examples of the exercise of 'creative political economy' – by both sovereign states and non-sovereign subnational island jurisdictions – include the sale of stamps and commemorative coins, Internet domain sites, such as the lucrative .tv top-level domain for Tuvalu (Aldrich and Connell, 1998, p. 76; Baldacchino and Mellor, 2015; Prasad, 2004); offering flags of convenience, as with Bermuda, Cyprus, Marshall Islands and Vanuatu (Powell, 2013); serving as *gulags* and detention centres that act as extra-territorial 'legal limbos', as with Guantánamo Bay in Cuba, Christmas Island and Nauru (Butler, 2002); as offshore finance centres, as with Bermuda, the British Virgin Islands and the Cayman Islands (Vlcek, 2008); as electronic-gaming hubs, as in the case of Antigua and Barbuda (Cooper, 2011); as 'Bitcoin island', as with the Isle of Man and, formerly, Alderney (Connell, 2014; Milmo, 2016); as casino tourism sites, as with Macau and Christmas Island (Wong and Siu, 2015); and similar, rent-based revenues (Baldacchino, 1993; Kakazu, 1994; Sidaway, 2007, p. 354).

These are all rich, contemporary examples that highlight the soft underbelly of sovereignty. Since the Treaties of Westphalia of 1648, the 'nation-state' has emerged as the organising political idea(l) of the modern age (Krasner, 1995). While states are inherently jealous of their territoriality, some of these very states have taken initiatives that have spawned subnational units that operate with some autonomy yet without full sovereignty. Much like the binary numeral system, sovereignty is often treated as indivisible. It is either 0 or 1: you either are sovereign or you are not, with hardly any room for dithering in between. But *this assumption is false*: it is part of the inherent 'governmentality' of states to challenge and test this 'yes/no' principle, seeking to exploit the many, lucrative policy possibilities and niches that it creates (Baldacchino, 2012a).

A history of playful sovereignty

Various special practices of sovereignty – legal niches in their own right – have emerged historically, and in their own turn have provided examples and precedents for similar practices of innovative governance.

A good place to start is the concept of *sanctuary*. The notion harks back to the Judaic Law of Moses, which held that fugitives from human laws could take refuge at the altar of God who, as the ultimate source of justice and mercy, would offer His protection if they were innocent. Later on, the Christian Church in Europe broadened the idea to include the blanket protection of the accused from the pursuit of human law. (The gypsy Esmeralda was granted sanctuary inside the Cathedral of Notre Dame in Paris in Victor Hugo's *The Hunchback of Notre Dame*.) As the political power of the Church waned, governments recovered their jurisdictional reach and ensured that any fugitive would be liable to pursuit and capture, even in holy places. Symbolic offerings of sanctity were offered to various draft dodgers during the US Vietnam war, with various members of the clergy appearing to be 'delighted with the opportunity to use their houses of worship in what they feel is an openly defiant way of supporting dissent' (*Time*, 1968). The application of Islamic Sharia law in otherwise liberal democratic polities today speaks to a similar 'carving out' of jurisdictional practice.

Second is the principle of *extra-territoriality*. Here one find a similar enclaving of jurisdiction, also involving chapels, but which persists and has become institutionalised in present-day international relations. This crafting of a jurisdictional enclave owes part of its origins to the so-called 'embassy chapel problem'. The needs of states to maintain open and ongoing diplomatic channels and relations often conflicted with the exigencies of national law: one effective way to beat that quandary, and in spite of the 'abuse' that such immunity begot (Trimble, 1946), was to designate an arrangement whereby the house and equipages of the diplomatic agent are regarded as the territory of the power by whom he is accredited. In what can be seen as an extension of the principle of sanctuary into the secular world, the agent of extraterritoriality – typically, a resident ambassador – is not a subject of the receiving government, and is thus exempt from local taxation and civil jurisdiction. For similar reasons, political or criminal offenders, and citizens who may feel that their civil rights are threatened, may and do seek asylum in a foreign embassy: a high-profile example of such behaviour would be WikiLeaks editor Julian Assange, who in 2012 sought refuge at the Embassy of Ecuador in London and was granted political asylum by Ecuador (Neuman and Ayala, 2012). This mechanism was an imaginative resolution to an otherwise thorny impasse, concerning in particular the religious practices of ambassadors post-sixteenth century. Thus, it would have been simply intolerable for the most Catholic king of Spain to condone the English ambassador participating in a mass according to the rites of the Anglican Church of England on Spanish soil; and just as impossible for the kings and queens of Britain, themselves head of the Anglican Church, to permit Spanish ambassadors to freely indulge in the 'papist' rites of the Catholic mass on English territory.

A third relevant historical development, and this time taking us literally offshore, concerns the practices of *piracy and corsairing*. Piracy had been a 'natural way of life' in the Mediterranean Sea since the tenth and eleventh centuries. Pisa, Genoa and Venice were the main maritime cities involved in the practice, while maintaining fleets to ward of foreign pirates in their own waters (Unger, 1980, p. 102). By the fourteenth century, however, a distinction in naval plundering emerged: corsairs emerged as entrepreneurs who would have some form of licence or permission from a parent state to commit aggressive acts. In principle, assaults were restricted to times of war and only against enemy vessels. But, human nature being what it is, these privateers were often not reliable 'business partners': they frequiently attacked friendly ships and were sometimes reluctant to share profits with their commissioners. The Knights Hospitaller of St John based in Malta, in particular, became well-known privatees in the sixteenth century and their targets were by no means confined to Muslim ships and states (Atauz, 2008). Thus, a risky but potentially lucrative economic activity emerged whereby the same action could be deemed legitimate or otherwise, depending on whether it was somehow sanctioned.

Fourth, and with another Venetian connection, is the concept of *quarantine*. This term, associated with health practices, owes its origins at least to 1377, when the Rector of the seaport of Ragusa, then belonging to the Venetian Republic, officially issued a 30-day isolation period for ships; a period that became 40 days for land travellers (*quaranta* means 40 in Italian). In 1423, Venice set up a *lazaretto* (quarantine station) on the island of Santa Maria di Nazareth near the city, and the Venetian system became a model for other countries to emulate (Gensini et al., 2004, p. 259; Sehdev, 2002). The zoning practice in this case was an act of containment, codified at law, to ensure that communicable diseases that may have been contracted from elsewhere – by sailors, passengers and their animals – would not infect and spread to local populations, with possibly lethal consequences. The surrounding sea made the possibility of escape, and thus widespread contagion, that much less likely. The ease of containment explains why many 'islands-as-exotopias' have hosted prisons (Alcatraz, Asinara, Devil's Island, Elba, Robben Island, St Helena), hospitals that dealt in leprosy or contagious diseases (e.g. the 'leper colonies' of D'Arcy Island, BC, Canada and Moloka'i, Hawai'i, USA) (Edmond and Smith, 2003); and now, more recently, immigrant detention centres (Mountz, 2011).

Fifth, moving into the twentieth century, and on the theme of economic activity, there is the emergence of *export processing zones* (EPZs). With business organisations developing their own legal personality, culture and behaviour patterns (Morgan, 1996), the export processing zone applies a similar logic to a whole cluster of firms. The legal regimen usually includes more relaxed labour and environmental laws, with preferential customs duties on selected import items, thus presenting an attractive proposition to foreign export-oriented capital. The EPZ becomes an enclave protected from the typically more stringent regulatory framework of the rest of the country: in some exceptional cases, and to prevent accusations of 'two weights, two measures', a whole island jurisdiction can transform itself into an EPZ, as Mauritius has done, with a thriving garment industry (Frankel, 2010). One of the first onshore EPZs was the duty-free Shannon International Airport in Ireland in 1959. Many airports remain duty-free enclaves today. But the full-scale, free *trade* export processing zone model was perfected with respect to Puerto Rico and Operation Bootstrap. The island's local administration enticed US-owned companies to relocate and set up operations on Puerto Rico by providing labour at costs below those on the mainland, access to US markets without import duties and profits that could enter the country free from federal taxation (Rivera-Batiz and Santiago, 1996). The investment of external capital, importing of raw material and the exporting of finished products to the US market were all supported by the state. Thanks to this clutch of measures, the island industrialised fast, overtaking for some time even the 'Asian tigers' in its staggering rate of growth. The policy started to stagnate in the early 1970s: a large public sector and a pervasive welfare state dampen motivation for enterprise (Goldsmith, 1979; *The Economist*, 2006).

Looking at these, similar and more recent experiments in innovative governance – like the flag of convenience, offshore casinos and electronic gam(b)ing – it is clear that islands are the premier geographies of choice for these forays. The notion of a sanctuary enisles a particular space; an overseas embassy effectively becomes a jurisdictional island, immune from most of the legal obligations of the surrounding host state. Throughout history, islands have been locales where both nature (via evolutionary processes) and humankind (via creative political economy) have undertaken many interesting experiments. This trend continues unabated.

Creative alternatives: beyond the 'zero-sum' solution

Pace examples like the Falklands/Malvinas, human history is well endowed with examples of islands – large and small, populated and uninhabited, single and archipelagic, in both warm and cold water locales – that have borne the brunt of territorial disputes and which have seen these resolved *not* in 'winner-takes-all' ways. Here are real-world examples of more interesting experiments like the ones just described, that defy the zero-sum approach, listed here in historical sequence, and which will be discussed in detail in the forthcoming chapters of this book:

- St Martin (France)/Sint Maarten (the Netherlands): co-habiting a divided island since 1648 (Hillebrink, 2012).
- A condominium with an alternating six-month/six-month governance: Pheasant Island (Spanish: *Isla de los Faisanes*, French: *Île des Faisans*, Basque: *Konpantzia*) is an uninhabited 2-acre islet on the Bidassoa river and the Franco-Spanish border. The island is under joint sovereignty of Spain and France, each of which controls the island for alternating periods of six months since the Treaty of the Pyrenees (1659) (Kliot and Newman, 2013, p. 270).
- Swapping: as happened in the case of the islands of Zanzibar for the island of Heligoland, ratified by the Anglo-German Agreement of 1890 (Drower, 2002).
- Vanuatu (known as New Hebrides before independence in 1980): governed as a condominium (co-/parallel governance by both Britain and France) for the period 1906–1980 (Rawlings, 2012).
- Svalbard: under Norwegian sovereignty; but an open territory for commercialisation to treaty signatories, since 1920 (Grydehøj et al., 2012).
- Åland: a demilitarised autonomy, with protected Swedish culture and language; yet part of the sovereign state of Finland since 1921 (Hepburn, 2014).
- Antarctica: a special regime, governed by its own treaty since 1959 (Dodds, 2011).
- Tuvalu (then, the Ellice Islands) and its secession from Kiribati (then, the Gilbert Islands) prior to the achievement of independence by both countries in 1978 and 1979 respectively (McIntyre, 2012).
- Picton, Lennox and Nueva islands, with mediation by the Pope: the islands and the three-mile territorial sea belong to Chile; while oceanic rights have been shared by both Argentina and Chile since 1984 (Guo, 2015).
- Shared jurisdiction: as with Ellis Island, USA, between the states of New York and New Jersey (since 1998). This arrangement proved to be a practical solution when the US Supreme Court found that the original island belonged to New York but the reclaimed land on the island belonged to New Jersey (Mottola, 1999).

- Shared exploitation of natural resources, without compromising claims to sovereignty (Valencia, 2007); this is the situation with the Timor Sea Treaty between Timor Leste and Australia over resources in the Timor Gap, since 2002 (Scofield, 2007).
- Bolshoi Ussuriiski/ Heixiazi: a 50–50 solution agreed to by Russia and China to conclude negotiations and stabilise their contiguous border along the Amur river, since 2004 (Iwashita, 2013).
- Reaffirmation of the disconnection between territorial waters and the definition of fishing zones in the agreement between the Bailiwick of Jersey and France, sealed in 2004, and an update of another landmark agreement, that of 1839. The terms of the agreement presented a complicated set of legal restrictions that not only included the number of vessels but also such other criteria as their port of origin, the size of the vessels; and a proscription against the number of vessels allowed to fish simultaneously. This agreement also defined a new type of marine space that integrated concepts and principles from newly evolved global legal frameworks with customary durability (that is, long-term use) (Fleury, 2011).
- Transboundary protected areas (TBPAs) as promoted by the International Union for the Conservation of Nature (IUCN)-World Conservation Union. A few of these are marine-based and involve islands, such as the Wadden Sea area, in Northern Europe. This is also a UNESCO World Heritage site since 2009; it comprises a string of islands involving Germany and the Netherlands, with Denmark also participating (Shine, 2007). Similarly, a marine park has been proposed for the Norman–Breton Gulf by the French government which could include the waters belonging to the Bailiwick of Jersey. At the early stages of the feasibility study for this project, it appears that Jersey authorities might not oppose the integration of their waters within its perimeter (Fleury, 2011, p. 40).
- Divided jurisdiction, as has been recently proposed for Hans Island/Tartupaluk (claimed by both Greenland/Denmark and Canada) (National Post, 2012).

Conclusion

These episodes deal with very different islands: from minuscule *Isla de los Faisanes*/*Île des Faisans* to continental Antarctica. But, throughout a span of almost 400 years, these islands have provided telling evidence of the possibility of workable solution protocols that do not assign islands summarily to one country, leaving other parties sulking, angry and licking wounds, contemplating revenge. Collectively, this string of episodes serves as a genealogy of creative sovereignty, facilitated no doubt by the circumscribed and finite nature of their island geography which discourages spill-overs.

It is perhaps to *these* episodes that those embroiled in the Diaoyu/Senkaku case should look for inspiration.

7 Splitting

Introduction

We start our review of solution protocols with the act of splitting. This involves the legal separation of what might, at the outset, appear as indissoluble territory into respective and separate jurisdictions. Where islands are concerned, splitting may involve carving out two island sub-set jurisdictions out of what had been hitherto seen as one archipelago; or splitting one single island into two jurisdictions. The rationale by which such a splitting can be organised can be based on claims of ethnic contiguity as in the case of Tuvalu from Kiribati; on claims of equity of land area, or the 50–50 solution, as in the case of Bolshoi Ussuriiskii/Heixiazi; or on the basis of a whimsical or banal exercise, as in the popular narrative explaining the splitting of St Martin. Details of all these cases, and how the splitting has come about, are provided below, penned by experts in their respective field. They appear in the chronological order of their jurisdictional split.

The splitting option is worth entertaining in the case of the Diaoyu/Senkaku. One is, after all, dealing here with an archipelago that comprises at least five islets and three sets of rocks. Moreover, Chiwei Yu/Taisho island and Diaoyu Dao/Uotsuri island, the largest members in the Diaoyu/Senkaku group, are separated by over 100 km of ocean. It is thus a mistake to think of the Diaoyu/Senkaku islands as a single, compact entity. As with many island territories, one may refer to an island but one would really mean and would actually be referring to a multi-island assemblage, an archipelago, which therefore makes splitting a distinct possibility. The act of splitting can also be described as the 'Solomonic solution', evocative of the story in the Hebrew Bible (1 Kings 3:16–28) whereby King Solomon proposed to split a live baby being claimed by two different mothers, thus tricking the two women into revealing their true feelings. Thus, even if somewhat macabre, a baby can be proposed to be split in two; it should be easier to propose a similar splitting of an island. Such an act has happened, and even amicably, as the St Martin case attests.

St Martin (Steven Hillebrink)

St Martin is the world's smallest inhabited and divided island. The French side, to the north, has a land area of 43 km^2 and a resident population of around 37,000; while the Dutch side, to the south, is smaller (at 34 km^2) but has a larger population, of 41,000. The island is a mass tourism destination, ranking highly on the tourism penetration index (McElroy and De Albuquerque, 1998): it is visited by around 1.5 million cruise ship passengers annually; and another 500,000 arrive every year via one of two international airports: the busier Princess Juliana in Dutch Sint Maarten (SMX); or L'Espérance, also known as Grand Case, in French Saint-Martin (SFG). The northern part is, since 2007, a separate overseas collectivity (*collectivité d'outre mer*) of France, with Marigot as its capital. The southern part is, since 2010, a separate 'country' (*land*) within the Kingdom of the Netherlands, with Philipsburg as its capital.

The island was discovered by Christopher Columbus on behalf of Spain in 1493; but its division dates from the seventeenth century. The Dutch began to settle in 1627, with the intention of exploiting the island's natural salt deposits. The Dutch erected a fort, which provoked a spirited attack from the Spanish, who wrested control in 1638. However, the Spanish deserted St Martin for good within a decade, considering it to be too small, of minimal strategic interest and unsuitable for plantation agriculture. By then, French settlers had started occupying the northeast coast of the island; while the Dutch West Indian Company occupied the area around the great salt pond in the south.

In 1648, the French–Dutch separation was formalised in the Treaty of Concordia, which provided that the French and Dutch shall live on the island as friends and allies. The Treaty recognised both the dual nationality and the unity of the island: there would be no physical border between the two nations and people and goods would move freely between the two zones. The original text of the Treaty of Concordia has been lost, perhaps when the British burned down the Court House in Philipsburg in 1810, disappointed that they did not find the treasures they had hoped for (Badejo, 2008, p. 5). But the contents (as reproduced in Tertre, 1667, pp. 412–14), have been recognised as binding by France and the Netherlands (Hoeneveld, 2002, pp. 90–1).

The Treaty instructed the French and Dutch authorities to determine the precise course of the border. A popular story on the island tells how the border is supposed to have been determined. A Frenchman and a Dutchman started walking in opposite directions from Oyster Pond. The Frenchman walked around the northern side of the island, the Dutchman along the southern side. At the place where they would meet up, they drew an imaginary line to Oyster Pond, which determined the border. The Dutchman walked slower – or took a nap halfway because he had too much to drink – which explains why the Dutch half is smaller. This story is used on the island to mock the Dutch, or to illustrate that the French are 'dangerous to the "relax and enjoy yourself" ideology of the island' (Guadeloupe, 2009, pp. 21–23).

Whether the walking contest has any relation to historic facts is unknown. The first written sources do not mention it. In the version of the French missionary Jean-Baptiste Tertre, the French and Dutch colonial authorities sent detachments to St Martin after the Spanish abandoned their occupation in 1648. The French and Dutch commanders agreed to revert to the situation that had briefly existed before the Spanish had chased them from the island (Tertre, 1667, pp. 409 et seq.). A later British version offers more details, apparently based on a visit to the island during the mid-eighteenth century (Jefferys, 1761). According to this account, French and Dutch prisoners of war, who had escaped to the woods during the Spanish occupation, agreed upon the division after the Spanish left. They built a canoe and the Dutchmen used it to inform the Dutch Governor in St Eustatius, from where the French authorities would also be informed. But 'from a true principle of Dutch perfidy, they forgot the last part of the errand', and returned to claim the entire island. The French authorities in St Kitts found out and sent troops, but their commander decided to offer a peaceful division of the island, which the Dutch accepted (Jefferys, 1761, pp. 75–76).

The formal separation of St Martin by way of a treaty seems to have been inspired by the division of the nearby island of St Kitts which was, mainly from 1627 to 1702, intermittently (but uneasily) split between the English and the French. This would also explain the treaty's reference to rivers and lakes, which did not exist on St Martin as a whole (Glasscock, 1985, p. 11). During the colonial period, it was not uncommon for Caribbean islands to be occupied by settlers from different European countries. Apart from St Kitts, other examples include Saint Croix (Dutch and British), Tobago (Latvian and Dutch) and Quisqueya, or Hispaniola (French and Spanish). All these separations, except the very last, have now been terminated; one of the reasons for this may have been that the early colonial economies shifted from trade (and piracy) to agriculture. The colonists thus shifted away from the coast and moved inland in search of arable land, where they then came into conflict with settlers from other countries.

St Martin did not provide much opportunity for agriculture, especially on the Dutch side: the island is relatively dry and unsuitable for the growing of many cash crops. This condition might explain why the island's division survived beyond the early colonial period, and why the island was not granted to the British in the Utrecht Peace Treaty of 1713 (Hartog, 1981, pp. 32–33). Nonetheless, there have been, over the centuries, intermittent drives towards the unification of the island: this would be done by military force, by buying up the other half, by petitioning the British Crown to take over the entire island, by proposing to sell the entire island to the United States, or by becoming an independent state (Speetjens, 2002). During the seventeenth and eighteenth centuries, the French, Spanish, British and Dutch on various occasions controlled the entire island, usually through military occupation when the island was not protected. Each time, the situation of the Treaty of Concordia was quickly – or after a few years – reinstated, as the European powers became less and less inclined to risk a military conflict on a Caribbean island with very little economic or strategic importance. During the nineteenth century, there were many French initiatives to acquire the Dutch side, and vice versa: for example, the Commander of the Dutch territory during the 1740s recommended that the West Indian Company should consider buying the French side to relieve the overpopulation being experienced on the Dutch side. He estimated that the French side could be bought for 150,000 Dutch guilders (Hartog, 1981, p. 41). Yet, none of these initiatives came to pass (Sicking, 2006). France – unlike Britain, the Netherlands and later the US – had never been in the habit of buying or exchanging colonies. Moreover, selling a colony became less and less attractive as Holland's economic and military power declined. From 1900, it was feared that the sale of one or more Dutch Caribbean islands would exacerbate a chain of events that would ultimately lead to the loss of Java, an island that was considered vital to the economic future of the Netherlands (Hillebrink and Bense, 2008).

After World War Two, the idea of uniting the island under a single colonial government became even more unrealistic, as the acquisition of a new territory in the Caribbean would clash with the international political drive for decolonisation that was so strong at the time. The idea of forming a single colony was replaced by the idea of becoming independent as a single country. Appeals for the independence of St Martin as a whole have been mooted, and this idea has gained some popularity since the 1980s, particularly on the Dutch side (Lake, 2000; Sekou et al., 1990); but it was never the subject of serious negotiations and was not put to the test of a referendum. On the contrary, plebiscites on both sides of the island have shown a clear preference by the population for continuing the constitutional ties with France and the Netherlands respectively (Baker, 2015; Hillebrink, 2008).

Described as 'an inseparable couple' (*Today*, 2012), the French and Dutch sides of the island have followed quite dissimilar political and economic trajectories since the 1950s. The French part of the island became an integral part of France and part of the European Union; the Dutch part became an autonomous country and is not part of the EU. The French part uses the euro as its currency, the Dutch part the US dollar. The French side enjoys the privileges of the French welfare state; Dutch social benefits do not extend to St Maarten. The island as a whole, however, has been transformed into a bustling regional hub, as well as one of the world's most densely populated territories. The recent changes in the constitutional status of the two sides make cross-border cooperation easier; the harmonisation of regulations and policies in such areas as taxation, the labour market, energy supply, tourism, roads and traffic is increasingly feasible, although the authorities so far have not made much use of the increased possibilities for cooperation and harmonisation. Meanwhile, some (especially French) policy makers are exasperated with the situation: 'In Dutch Sint Maarten, the rules are more relaxed, salaries and taxes are lower and products are cheaper' (*Today*, 2014). No wonder: local entrepreneurs, residents and tourists, many of whom are neither French nor Dutch, shrewdly choose to reside, work, visit or pay taxes where conditions on the island are deemed more favourable.

Dividing a single and densely populated island like St Martin is one story; dividing a sprawling and lightly populated archipelago is quite another. We move from the Caribbean to the Pacific to consider the partition of the Ellice Islands (modern day Tuvalu) from the Gilbert Islands (modern day Kiribati).

Tuvalu (David McIntyre)

By both land surface area and population, Tuvalu is the world's smallest and most isolated independent archipelago state. When it secured political independence from the United Kingdom in 1978, it was the world's third poorest country. Its population of 11,000 resides on a land area that, collectively, amounts to just 25.6 km^2 (9.9 square miles); while its exclusive economic zone (EEZ) covers a huge oceanic area of some 700,000 km^2 (270,000 square miles). The country consists of three reef islands (one of which was uninhabited until the late nineteenth century) and six atolls, situated along a 600 km (373 miles) roughly north–south axis; each of the atolls consists of various islands and islets. Its capital is located on Funafuti atoll (on Fogafale Islet), where over half the population resides. The closest inhabited land to Tuvalu is Fiji, some 1,180 km (730 miles) away to the south.

With such a forlorn location, a minuscule and dispersed population and a scant (terrestrial) natural resource base, Tuvalu's prospects for 'economic development' in any conventional sense of the term are loaded with major long-term challenges. For a number of reasons, the country is recognised as one of the world's most environmentally vulnerable island nations: a long and severe drought led to an official state of emergency and the frantic delivery of desalinators and fresh water supplies by air and sea (Perry, 2011). The country is also facing the unappealing prospect of wholesale disappearance below the surface of the ocean, a likely consequence of climate change and sea-level rise (Farbotko and Lazrus, 2012). That the country continues to survive fairly well is mainly due to revenue from the lease of fishing rights in its extensive EEZ, its shipping registry, income disbursements accruing from the Tuvalu Trust Fund, aid project funding and cash grants, taxes, remittances and the lease of .tv: its country-code top-level domain name (Baldacchino and Mellor, 2015).

Why and how, then, did Tuvalu – then known as the Ellice Islands – opt to separate from the relatively larger Gilbert Islands – modern-day Kiribati, a sovereign, and itself a small, archipelago state – and risk its destiny as an independent country?

The Gilbert and Ellice Islands Colony (GEIC) lay astride the equator in the central Pacific between longitude 150°W and 170°W, spread over 5 million km^2 of ocean. The sprawling group was proclaimed a British Protectorate in 1892 (Munro and Firth, 1986, pp. 63–67). Increased German interest in the Pacific region and the outbreak of World War One speeded up its conversion to a Crown Colony in 1916; Banaba (Ocean Island) and the northern Line Islands were also added; while Kiritimati (Christmas Island), the most northerly of the Line Islands, was added in 1919; and finally eight of the Phoenix Islands were added, in 1937 (Maude, 1968, pp. 315–342). In these ways, the GEIC became a motley and very scattered incremental dependency, the legacy of a complex history.

Japanese forces occupied parts of the Gilberts in late 1941, the western Pacific limit of the Japanese empire. US forces took control of the Ellice Islands in late 1942, and constructed various runways and military bases, pre-empting Japan's southward advance. The Ellice atolls of Funafuti, Nukufetau and Nanumea were staging posts, their lagoons suitable for shipping and seaplanes, and their small flat geography ideal for airbases. These sites were used for the run-up to the battles of Tarawa and Makin in the Gilberts in November 1943, and the launch of the US offensive push northward through the central Pacific.

The partition of the GEIC into two, separate, dependent territories occurred in 1975. The Ellice Islands, renamed Tuvalu (meaning 'eight standing together', after its eight populated islands and atolls), went on to secure independence three years later, with a population of not more than 7,000. The Gilbert Islands followed suit in 1979 as the Republic of Kiribati (pronounced Ki-ri-bass), with a population of 53,000.

Such fragmentation of already small populations went decidedly against the grain of British administrative practice, United Nations thinking and economic development precepts: very small territories were deemed politically and economically non-viable. Fragmentation was also frowned upon, even if this meant sticking to artificial borders that had been put into place during colonial rule. Nevertheless, the official Whitehall mantra that dependencies must not be forced into constitutional arrangements they did not want eventually prevailed. So, after the Ellice Islanders had expressed themselves decisively in an October 1974 referendum (88 per cent turnout; 92 per cent in favour of separation) observed by a UN Visiting Mission (Chambers and Chambers, 2000, pp. 8–9), the British gave way and organised the creation of the newly dependent territory of Tuvalu, while at the same time hoping to ensure that this would not prejudice the future of the rest of the GEIC.

The main difference between the Gilbertese and Ellice islanders has been ethnicity, reinforced by colonial practices. All but one of the Ellice Islands were peopled by Polynesians; and even on Nui, the exception, people consider themselves 100 per cent Tuvaluan, retain some Gilbertese social and cultural traditions and speak a distinct Nui dialect which is a mix of the Gilbertese and Tuvaluan languages. The darker-skinned Gilbertese and Banabans were Micronesian. The two groups spoke different languages. The Ellice Islanders, of Tongan and Samoan origin, were very status conscious, upholding traditional social structures in which hereditary *aliki* (chiefs) were recognised as leaders whose authority was respected. Then, with the coming of the London Missionary Society (LMS) missionaries from the 1860s onwards, new sources of authority were introduced. Samoan pastors of the LMS came to rival the *aliki* in power and prestige (Munro, 1996, pp. 124–157; Laracy, 1983, pp. 19–24). In the Gilberts, the hierarchical principle was tempered by a system of rule by councils of elders, the *unimane*, who gathered in village meeting houses, *maneaba*, where careful mediation over the allocation of the limited land enabled the people to survive in an environment with scarce natural resources (Macdonald, 1982, pp. 6–10).

For much of the colonial period, the two groups of islanders had little contact. They were under one government but nothing was done to foster a common sense of identity. By administering the different cultural groups as one administratively convenient entity, regulations framed initially for the more numerous Gilbertese were inappropriately applied to the Ellice Islanders. Then, with the development of educational provision, the Ellice Islanders won disproportionately more scholarships and gained more positions in the colony's public service relative to their population numbers. Political stirrings in the 1960s arose through Gilbertese resentment over the alleged Polynesian superiority of their Ellice colleagues (Butcher, 2012, pp. 28–31; Macdonald, 1982, pp. 134–138).

In the early 1970s, new constitutional arrangements exposed the Gilbertese and Ellice antagonisms. With greater political representation, the Ellice Islanders feared a tyranny of the majority: they concluded that they risked becoming a neglected minority group and losing their identity; they were also anxious that they would be effectively shut out of employment opportunities by the Gilbertese in any post-independence arrangement. They therefore demanded separation. If they did not separate, 'the Ellice identity will be lost into, or swallowed up by the Gilberts within another 20 years or so' because of inter-marriage and widespread use of spoken Gilbertese, commented a Funafuti resident (McIntyre, 2014, p. 205).

Britain complied. Tuvalu came into being on 1 October 1975. On 1 January 1976, the separate administration took over at Funafuti. In December 1976, a motion for independence was successfully moved. Independence Day was 1 October 1978, the third anniversary of separation. Kiribati independence followed on 12 July 1979.

Each country has survived; but, since the Tuvaluans and I-Kiribati had separate identities, and aspects of their cultures and values were different, their split may have seemed unlikely, but was somewhat inevitable. The GEIC government had never endeavoured to forge a nation and Tuvalu would not stand for permanent minority status. Ideas of federation, loosely floated in the 1950s and 1960s, were never seriously pursued. Tuvaluans chose the difficult path 'to preserve their cultural identity' (Macdonald, 1975, p. 43) which, however, seems to have paid off, at least for now: Tuvalu's gross national income (GNI) per capita is over US$3,400 almost twice that for Kiribati, at US$2,010 (2012 estimates) (Howe, 2013; UNFPA, 2014).

From the split in a sprawling archipelago that led to the birth of what are now two sovereign states, we move to a different splitting exercise, one that concerns two river islands with scant populations that happen to lie along the Russian–Chinese border delineated by the Amur River.

Bolshoi Ussuriiskii, Russia/Heixiazi, PRC (Akihiro Iwashita)

A large delta between the Ussuri and Amur rivers (in Russian) or the Heilongjiang and Wusuli rivers (in Mandarin Chinese) has been the locus of a series of serious challenges between China and Russia (or the former Soviet Union) for centuries. The delta is almost 350 km^2 in area and is located very close to Khabarovsk, the second largest city of the Russian Far East (after Vladivostok). The delta includes two sparsely inhabited islands, called Bolshoi Ussuriiskii (óстров Большóй Уссури́йский) and smaller Tarabarov (óстров Тарабаров as known in Russian) or Heixiazi Dao (黑瞎子島) and Yinlong Dao (銀龍島) respectively (in Mandarin). This remote delta area had been occupied by Russia since 1929, but China had maintained its claims to this territory. Deadlock over the future of this region has been the main stumbling block to the full resolution of Sino-Soviet border quarrels for various decades. The dispute led to the Zhenbao/Damanskii military confrontation in 1969 and drove the two neighbouring countries to the brink of a nuclear war. Since 2008, however, a 'fifty–fifty' solution has been agreed to by both sides and implemented. Beijing and Moscow are now proudly showcasing the delta as a working example of 'one island, two countries'; while the citizens of both countries, and particularly those inhabiting the affected border regions, are now able to securely develop a trans-national cross-border system of economic and cultural cooperation (Iwashita, 2013).

Russian and Chinese historians do not agree on the background conditions that led to their respective claims to the area. And Moscow never recognised the inequality of the relevant past treaties which, Beijing claims, were imposed unfairly on China by Russia in the nineteenth century.

Under the Qing Dynasty, China was engaged in and distracted by its Second Opium War, and was being humiliatingly pushed back from the coastal regions around the Sea of Japan. Meanwhile, after the Aigun Treaty (1858) and the Peking Treaty (1860), an emboldened Russian empire took control over the Amur and Ussuri rivers and extended its power towards the Sea of Japan and the Korean Peninsula. These treaties reversed the outcome of the Treaty of Nerchinsk (1689) in Russia`s favour, and established what remains today as the generally disposition of the long Sino-Russian border. China, however, claimed these to be 'unequal treaties'; it has since continued to claim the ceded lands at this delta region without fail. To no avail: Russia completely controlled the islands and the surrounding channels around the Amur–Ussuri delta, especially after the so-called Chinese Eastern Railway Incident (1929), where the Soviet Union reversed a Chinese incursion into a complete Russian victory (Patrikeeff, 2002).

Sino-Russian border negotiations opened in 1963–1964. A Russian proposal to swap some islands was met by a China adamant to share the Amur–Ussuri river delta. Since this proposal was intended as part of an all-inclusive package deal, negotiations struck deadlock. The 7,000-plus km Sino-Soviet border, from Mongolia to North Korea – the longest international border between two countries in the world – became a dangerous non-regulated line, and border guards from both countries were forced to watch over it with heightened tension.

It was here that the frustrated Chinese sought to break the political stalemate. In March 1969, China launched various surprise military incursions over the border, including an attack on Russian soldiers patrolling Damanskii Island (Остров Даманский), known in Chinese as Zhenbao Island (珍宝岛) on the Ussuri river. The Chinese eventually retreated; the two sides reported different numbers of casualties (Goldstein, 2001; Riabushkin, 2006).

Full-scale negotiations over the border issue were resumed only after Gorbachev indicated, in 1986, the USSR's willingness to establish cordial and long-lasting peaceful relations with its neighbours (Lapidus, 1987).

Most delimitation issues on the Sino-Soviet border were finally resolved after a couple of tough negotiation rounds (Kireev, 2006). The islands along the delta near Khabarovsk between the Amur and the Ussuri rivers remained, however, 'a thorn in the side of peaceful Sino-Russian relations' (Levenstein, 2011, n.p.). Agreement had been reached on some 98 per cent of the disputed territory (Iwashita, 2009). The 1991 agreement on the Sino-Soviet eastern boundary was conceived in this way by excluding the fate of the Amur–Ussuri delta to future negotiations. This agreement reflected the seriousness of the challenges, and highlighted the strategic and symbolic importance of the disputed delta region to both sides; but it also paved the way to an eventual solution, thanks to considerable confidence building (Hyer, 1996).

In 2001, when Russia and China were preparing to sign a 20-year Treaty of Good-neighbourliness, Friendship and Cooperation, Moscow seemed eager to have all remaining problems on the territorial issue resolved in time for the signing of the treaty (Ministry of Foreign Affairs of the PRC, 2001). It is alleged that Russian President Vladimir Putin hoped to hand over Abagaitui island (land area: 58km^2) on the Argun river to China, in return for keeping the delta between the Amur and Ussuri rivers under Russia's complete control. China, once again, is said to have rejected Russia's proposal (Iwashita, 2013).

Many expected the stalemate to drag on; but a surprise was in store. A high-level summit meeting between China and Russia on 14 October 2004 put a sudden end to all remaining border disputes, including the lingering delta question. The solution that had been reached was leaked to the public just after Russian President Vladimir Putin and General Secretary of the Chinese Communist Party Hu Jintao had declared a historic 'win–win' resolution for finalising all pending territorial problems: all remaining disputed islands would be divided between Russia and China on the basis of a clinical, 'fifty–fifty' solution, based on land area. As far as Bolshoi Ussuriiskii and Tarabarov islands on the Amur river delta were concerned, they would be partitioned in almost equal halves: Tarabarov (some 40 km^2) was handed over to China in its entirety; but Bolshoi Ussuriiskii (some 300 km^2) would be partitioned off such that, in total, both countries each ended up with some 160–170 km^2 of the delta in all. That meant that some 120–130 km^2 of Bolshoi Ussuriiskii were handed over to China; while the remaining 160–170 km^2 remained as part of Russia. (The exact size of the land area remains unclear; it is subject to continual erosion and silt deposition.) With this agreement, Russia, the world's largest country by land area, got to be just a little smaller (RT, 2008).

Admittedly, both sides continued a round of secret negotiations for dealing with these last-remaining, disputed islands. It seems that every imaginable attempt was being made to prevent the partition of small islands. The

Khabarovsk elite was not keen on sharing the delta with China, and feared Chinese infiltration; it lobbied vigorously against partition ((Iwashita, 2013). Russia even proposed to provide China with some surplus destroyers in exchange for the whole of Heixiazi Island (Iwashita, 2005a). China refused, and in turn, suggested it would give up Heixiazi in exchange for some territory around the Tumen River, which would provide access to the Sea of Japan. It was Russia's turn to decline the offer.

The '50–50' policy was clearly a last resort, but a possible resort nevertheless. Both sides had avoided splitting whole islands, preferring instead to maintain a single and unambiguous management and governance structure even for what were very small island spaces. This 'Judgment of Solomon' type solution had, however, been tried before: in 1997, when China and Russia had agreed to divide a disputed area of some 300 hectares in the Khazan region, near the trilateral Sino-Russia–Korean border: 140 of these hectares remained Russian; and 160 hectares moved to China (Iwashita, 2007). That probably represented the first case of the 'fifty–fifty' formula being used to resolve a land dispute through a political judgement. Once such a precedent had been set, similar cases followed in other border disputes involving parts of the former USSR: these include the 1998 Sino-Kazakh agreement, the 1999 Sino-Kyrgyz agreement and the 2002 Sino-Tajik agreement. These were pragmatic departures from a 'winner takes all' strategy that was clearly unacceptable to one of the parties (Maxwell, 2007).

In July 2008, the two sides signed the protocol for the 2004 agreement and finalised all necessary procedures that made the division of the islands administratively possible (Weitz, 2008). On 14 October 2008 – exactly four years to the day of the agreement – China and Russia celebrated the opening of new boundary makers on the delta with a formal ceremony (*The Economist*, 2008).

The case of Bolshoi Ussuriiski/Heixiazi is the only known case of a populated, albeit small, island divided between two countries on the basis of an (almost) equal division of land area (rather than, for example, on the basis of geographic expediency).

The novelty of the situation goes further: in 2010, China and Russia issued a joint programme for the mutual development of the delta. The two separate plans look similar: the goal is to enhance local economic activity, set up a tourist and recreation zone, and guarantee the rich ecological diversity on the delta islands. Both also agree to open a border checkpoint and promote the free entry and exit of personnel within the islands as if they were unified with no barriers (Iwashita, 2004). Ongoing development plans for the border islands include conservation areas and buffer zones, military facilities, a security tower, customs office, and shopping and rest areas for visitors. The joint promotion of 'the entire China–Russia border in this area as a tourist and business hub' is notionally under discussion (*China Military News*, 2010). Long but sustainable spells of collaboration between different levels of government of the two neighbouring countries with a long history of conflict would accomplish what is effectively a ground-breaking experiment in Sino-Russian relations.

Thanks to the '50–50' compromise, China and Russia have secured more stable cross-border relations; the climate is certainly much less edgy than in 1969. Both countries have nurtured trust and built confidence in the process of securing this achievement, and there is relative peace of mind along the border to vouch for it.

Conclusion

The carving out of two (or more) jurisdictions from the same island, or set of islands, is a practical solution that is worth keeping in mind, especially when dealing with an archipelago as is the Diaoyu/Senkaku chain. But there are many other ways in which one can skin the proverbial cat. The next chapter, titled 'sharing', examines situations of condominia where multiple governments – national or subnational – may exercise authority concurrently on and for the same island, but – unlike the splitting described in this chapter – jurisdiction is not carved out in material or geographical terms. Instead, the rationale for the sharing could be ethnic/linguistic background (as in the case of the former New Hebrides), switching between two jurisdictions via a fixed time period (Pheasant Island, Spain/France), or simply being forced to suffer joint governance out of sheer practical necessity (Ellis Island, USA).

8 Sharing

Introduction

Despite their many differences, and as the Falklands/Malvinas case illustrates, boundary disputes have been strikingly consistent in at least one respect: they have tended to defy durable, sustainable solutions. Politicians, international law jurists, negotiators and scholars have largely failed to successfully craft solutions for the many complex problems created by boundary disputes, sticking in the main to a determination of who is right and who is wrong. In light of such failure, the time may have come to seriously revisit a 'largely abandoned theory of boundary dispute resolution': the condominium (Samuels, 2007, p. 728). Its rarity in application stems not from its lack of success, but rather from a continued obsession to resort to dogmatism in lieu of pragmatism (Jacobs, 2012, n.p.). After all, and paraphrasing Judge Huber in his *Island of Palmas* judgment, territorial sovereignty belongs always to one state, *but in exceptional circumstances to several states*, to the exclusion of all other states (my emphasis) (Post, 2016).

The foundations of the condominium concept can be traced to Roman law and its rules about undivided joint property, or *communio pro indiviso* (Van Der Merwe, 2015, pp. 9–24). Indeed, the best known usage of the term *condominium* today is far removed from issues of political governance. The 'condo' is typically a legal form of housing tenure in many countries where a specified section of a building (usually an apartment unit or flat) is individually owned, while access to common facilities in the same building – such as hallways, heating systems, elevators, garages, swimming pools and exterior areas – are controlled by the association of owners that jointly represent ownership of the whole piece. The concept, however, also has a different and less well-known history and usage in international relations, and specifically as a mechanism for conflict resolution. It may yet discover a fresh relevance in relation to the resolution of disputes similar to the Diaoyu/Senkaku matter. In fact, Japan has some experience in dealing with the concept, and involving an island, in the nineteenth century: Russia and Japan handled Sakhalin Island as a condominium, starting with the Treaty of Shimoda in 1855 and lasting until 1875, when Japan ceded sovereignty over Sakhalin to Russia (Nester, 1993).

Meanwhile, cases of 'interim condominia' – with contending parties striking pragmatic deals of co-administration while claiming to be shelving their claims to sovereignty, but permitting cross-border cooperation for, say, joint exploration and resource exploitation (about which more in Chapter 11), as well as pollution control or ecosystem conservation – are on the increase. Interim arrangements might very well turn out to be open-ended; and thus, for most intents and purposes, effectively permanent.

A *condominium* is thus also a political territory over which typically two (but hypothetically even more) sovereign powers (usually states, but also sub-national jurisdictions, and rarely individuals) formally agree to share *dominium* (in the sense of sovereignty) equally and to exercise their rights jointly, without splitting or dividing the territory up into discrete, 'national' chunks or zones.

Water-based condominia

Canton and Enderbury atolls

Some underlying agreements concoct condominia that can delineate particular rights and powers in favour of just one sovereign power: sovereignty can be divided and partitioned by one state in favour of another. One example of this relates to Canton and Enderbury atoll islands, placed in a 50-year US–UK condominium in 1939, but which joined Kiribati when the latter became an independent state in 1979. Here, the (now defunct) US airline Pan American was granted the rights to build an airport, and the British were to have access to the facilities for a reasonable fee (Megaw, 1977).

Germany–Luxembourg border

One of the oldest, and extant, instances of inter-country condominia is to be found along the acquatic border between Germany and Luxembourg. Borders between sovereign states usually run down the middle of water bodies; but, since 1815, the bed and banks of the rivers Moselle, Our and Sauer belong, by treaty, to *both* Germany and Luxembourg: it was the Treaty of Vienna that stipulated that 'the rivers themselves, where they form the border, belong in common to the two neighbouring powers [Prussia and the Netherlands at the time]'. The countries also share bridges and at least the tip of one uninhabited islet, Staustufe Apach, near the Luxembourg town of Schengen where, in 1985, several European countries decided to abolish internal border controls (the rest of the islet is French) (Smith, 1994, p. 21). In these few shared spaces of land and sea, one would be in both countries *at the same time* (Jacobs, 2012).

Germany–Netherlands border

The Ems-Dollart (or Eems-Dollard) Estuary functions as an 'implicit condominium' along the border between Germany and the Netherlands (Post, 2016). Here, there is no formal agreement in place; but the *de facto* situation is for Germany and the Netherlands to have decided to exercise their sovereignty jointly.

The Wadden Sea, which includes this estuary, is subsumed under the 1971 RAMSAR wetlands convention and has been, since 2009, inscribed as a UNESCO World Heritage Site. This area is governed by its own impressive slate of trans-boundary legal, policy and management tools: more about this in Chapter 13.

Brazil–Paraguay border

Another aquatic condominium that straddles national borders concerns the 200 km stretch of waters of the Paraná river, shared between Brazil and Paraguay, from and including the Salto Grande de Sete Quedas (or Salto del Guaira), to the mouth of the Iguaçu River: an agreement sealed via a 1973 Treaty (Nickson, 1982; Silk and Ciruna, 2013, p. 93). This is an example of a condominium set up to manage trans-boundary water resources by two neighbouring states (Leb, 2013).

Gulf of Fonseca

Another water-related case deals with the Gulf of Fonseca where three Central American countries – Honduras, El Salvador and Nicaragua – each with a coastline along the Gulf have been involved in a lengthy dispute over the rights to the Gulf and the islands located within. In 1992, the ICJ decided the case; allocating single sovereignty to the disputed islands (see also Chapter 1), but determining that the three countries are to share control of the Gulf of Fonseca itself, effectively creating an aquatic *tridominium*. The only stretch of water in the Gulf that was to be maintained as sovereign was the three-mile zone off the coastline of each riparian state (ICJ, 1992).

This last case is of particular interest to the Diaoyu/Senkaku affair, and for various reasons. First, it was a workable solution protocol created, not by express agreement among the concerned parties, but rather by judicial decision. Indeed, the condominium option was imposed by the ICJ in the Gulf of Fonseca case for the first time (Harris, 2012, p. 611). Second, it concerned neighbouring states that shared a history of skirmishes, troop movements and other military build-ups at their respective borders: a practice meant largely to intimidate the other, while keeping the conflict alive and festering (Evans and Shaw, 1993). Third, although the dispute was in principle one of sovereign control, resource development is a critical intervening variable: the area is believed to hold rich fishing grounds, as well as natural gas and oil resources. Fourth, this was the first case in which a third party was allowed to (partly) participate in the proceedings of the ICJ.

Lake Constance

Finally, Lake Constance, which borders Switzerland, Germany and Austria, is another *de facto* tridominium: it is considered to be legally so only by Austria. It is the only area in Europe where no borders exist, because there is no legally binding agreement as to where the borders lie; the countries involved have not been too keen in pursuing the case (Blatter, 2001; Jennings, 2014). However, this has not hindered Lake Constance from becoming the site of one of the world's first cross-border water conservation programmes; the *Umweltrat Bodensee* (Environmental Council of Lake Constance) wields substantial heft in its trilateral administration (Blatter, 2004).

Land-based condominia (non-islands)

Where shared sovereignty over land is concerned, examples are rarer, but are still more common than one might think. There was an Anglo-French condominium experience in Cameroon, which lasted only for two years (1914–1916) (Elango, 1987); while an Anglo-Egyptian hybrid prevailed over the Sudan from 1899 to 1955, even though Egypt was itself a British protectorate, in fact or at law, until 1922 (Sharkey, 2003). Other versions of the condominium arrangement prevailed over Samoa (1889–1899) with tripartite rule by Germany, UK and US; and the Free Territory of Trieste (1947–1954) involving Italy and Yugoslavia (Samuels, 2007, pp. 745–752).

Andorra

It may not be an island, but it has qualities of being enisled in the mountainous Pyrenees. A suitable opening example and one of the best known would be Andorra. Today, this is a small, independent parliamentary democracy. But, for over seven centuries (1278 to 1993), it was a co-principality of the Kingdom (and then state) of France and the Bishop of Urgell, Spain (Perkins, 2014).

Moresnet

A lesser-known case is that of Moresnet, a Dutch (and then Belgian after 1830)–Prussian demilitarised and landlocked condominium from 1816 to 1919, following the impossibility on reaching agreement about proper title in the 1815 Act of the Congress of Vienna, and absorbed wholly into Belgium with the Treaty of Versailles after WWI (Press, 2010).

Land-based condominia (islands)

Ellis Island (USA)

Possibly the most symbolic case of all is the condominium arrangement hoisted on the US neighbouring states of New York (NY) and New Jersey (NJ) with regards to Ellis Island, the site of the first landing for about 12 million immigrants to the United States. The island – now with a land area of 27.5 acres (11.1 hectares) – was the country's key national gateway and busiest immigrant inspection station from 1892 until 1954, during the heyday of long-distance sea travel (Chermayeff et al., 1991). The island was acquired by a Samuel Ellis, a colonial New Yorker possibly from Wales, around the time of the American Revolution, hence the name.

It has been owned and administered by the US federal government since 1808; has been operated by the National Park Service since 1965, when it was incorporated with the Statue of Liberty National Monument, and has hosted a museum of immigration since 1990.

After the British took over New Amsterdam (present day city of New York) in 1664, the colonial treaty provided for the Hudson River as the demarcation of the NY–NJ border, rather than the more common definition of the border at the midpoint of the river. The aggressive assertion of territorial rights by New York's fourth colonial governor, Edmund Andros, further ensured that all the southern islands in NY harbour, including those on the NJ side of the river (such as Liberty, Governors and Staten islands) were within the jurisdiction of NY. The river midpoint as boundary was finally set in 1834, but the islands were not affected by this ruling and remaining ensconced within NY.

The US federal government began expanding the island by landfill to accommodate the immigration station that opened in 1892. Landfilling continued in various stages until 1934. With this long period of land reclamation, some 24.5 acres were added to the original island. As a result, Ellis Island grew to ten times its original size. NJ contended that the new extensions were part of New Jersey, since they were not part of the original island; and in 1997, NJ filed suit to establish its jurisdiction. The dispute made it to the US Supreme Court (1998): this ruled that NJ had jurisdiction over all portions of the island created after the original compact was approved (effectively, more than 80 per cent of the island's present land), since the aggregate used for the landfill had been quarried and sourced from NJ. As a result, NY was granted jurisdiction to 4.68 acres; an area that includes almost all the main immigration building plus sections of the Wall of Honour, baggage handling/dormitory building, power house and administration building. This NY portion of Ellis Island is, however, effectively landlocked, enclaved within NJ's territory, which comprises the remaining 22.80 acres of the island. These include portions of the above buildings, the ferry house, and all the 29 hospital and dormitory buildings located on the island's southern flank (Castagna et al., 1999; Greenhouse, 1998). This verdict caused some anxious episodes: some buildings on the island, for instance, fell into the territory of both NJ and NY. Out of practical consideration, convenience, fairness, and perhaps some exasperation, NY and NJ soon agreed to share jurisdiction of the island (Hoskins, 2012).

Perhaps the best-known and longest condominium experience over inhabited territory concerned the whole country of the New Hebrides, which is the pre-independence name of the Republic of Vanuatu.

Vanuatu (David McIntyre)

An archipelago state, Vanuatu consists of a chain of 13 main and various smaller islands located in the southwest Pacific, some 800 km (500 miles) west of the Fiji Islands and 1,700 km (1,100 miles) east of Australia. The islands extend in a northwest to southeast direction for some 650 km (400 miles) in a rough Y shape. With a resident population of around 255,000 ni-Vanuatu (as the locals are called), the country is officially a small island developing state. The name Vanuatu translates as 'our land forever' in most of the local Melanesian languages of which there are hundreds: Vanuatu is considered to be the country with the world's highest density of languages per capita, with a mean of just about 2,000 speakers for each indigenous language (François, 2012). The country has three official languages: English, French and Bislama (a creole tongue). The capital, largest city and commercial centre is Port Vila (or simply Vila), on the island of Éfaté, with around 30,000 residents. The Republic of Vanuatu secured its independence in 1980.

Prior to independence, Vanuatu was known as the New Hebrides, a name given by British explorer Captain James Cook in 1774. European contact had begun earlier, with De Quirós on behalf of Spain (1606) and French navigator De Bougainville (1768), but the islanders were not keen on petty exchange with foreigners and developed a reputation for being inhospitable, savage (and cannibalistic) (Brunton, 1997). Then, starting in the nineteenth century, the islands were visited more regularly by Europeans wishing to trade, preach and convert the locals to Christianity, or to settle permanently amongst them as planters or businesspersons (van Trease, 1987, p. 2). The main commercial interest was in sandalwood (*Santalum austrocaledonicum*) which remains an important export crop (Page et al., 2010, 2012). Significant cultural change occurred only after the 1860s, when thousands of ni-Vanuatu who had been previously indentured to work on plantations in Fiji, New Caledonia and Queensland, Australia, began to return home. Many established new forms of political influence within the nexus of Protestant (mainly Presbyterian) missions, or successfully competed against European traders and planters settled on the islands. To safeguard the interests of the (mainly British) missionaries and (mainly French) planters, both the British and French governments established rudimentary political control over the islands with a Joint Naval Commission in 1887. This was set up initially for the sole purpose of protecting French and British citizens, yet claimed no jurisdiction over internal native affairs. After the rogue municipality of Franceville (present-day Port Vila) seceded and declared independence for a few months in 1889, however, the Naval Commission was replaced by a more structured British–French condominium in 1906 (O'Connell, 1970). The jointly administered Anglo-French condominium of the New Hebrides persevered until independence in 1980.

The principles of this condominium were laid out in the General Instructions to the British and French High Commissioners (acting through the respective resident Commissioners in Vila) of 29 August1907. These instructions prescribe the essence of a joint/shared sovereignty arrangement:

> The desire of the two governments is to secure the exercise of their paramount rights in the New Hebrides. The two Powers, who were mutually bound not to intervene separately in the New Hebrides, now agree to intervene there together. Instead of remaining mutually exclusive, their paramount rights are combined; the two countries jointly assume jurisdiction in the islands, and thereby provide against the possible appearance of a third Power.
>
> (Brant and Maycock, 1911, p. 519)

Officers of both nationalities staffed the administrative departments, and actions by the departments were subject to review by the resident Commissioners acting jointly. French and British Condominium agents were deployed in the four national districts and were granted joint control over their respective regions.

Consider the case of the police force as an example of this condominium in action. From 1906, the New Hebrides police force was divided into two separate contingents of armed native officers: one British; one French. The two forces reported to two national commandants, each of whom served under his (these were all men) resident commissioner. Each force was responsible for policing the actions of its own nationals. The cost of maintaining each police force was covered by its own national government, except in the case of joint operations, in which case expenses were paid from condominium funds (themselves provided by block grants from London and Paris) (Robson, 1963).

With the condominium arrangement, resident commissioners in Port Vila retained responsibility over their own nationals and jointly ruled the indigenous people; although this administrative procedure had minimal impact on most ni-Vanuatu, whose chief European contact continued to be with either (mainly British) missionaries or (mainly French) plantocrats. French subjects formed the land-owning and commercial classes, as well as most of the urban population: they held economic sway in the islands, seeking to extend their influence in the rural areas through the Catholic Church and by supporting the local pagan chiefs. Meanwhile, the British, who would never have been there in the first place were it not for the goading of the Presbyterian Mission, came to represent the aspirations of the Protestant flock and rural folk, amounting to over two-thirds of the population (Foster, 2008, pp. 369–370).

The islands became a major Allied base during World War II, when the spectacle of free-spending US troops inspired the transformation of the Jon (or John) Frum cargo cult on the southern island of Tanna into a significant anti-European political movement (Worsley, 1957). Espiritu Santo, the largest island, became the second largest US base in the Pacific, after Honolulu, with more than 40,000 troops stationed on the island at any one time. After the war, local political initiatives originated in concerns over land ownership. At that time, over one-third of land in the New Hebrides was owned by foreigners. The country is best known today mainly for its appeal as a green/wreck-diving tourism destination (Trau, 2012); and for topping the Happy Planet Index – based on life expectancy, human well-being and damage done via a country's environmental footprint – when this was inaugurated in 2006 (Campbell, 2006).

Joint rule is a 'hybrid form of government' (Daly, 2003, p. 1), a political expediency that persisted in the New Hebrides for 74 years, and in a few other countries for less time.

Some observers claim that the condominium device did not prove successful in practice: as an experiment in political administration, the New Hebrides condominium has been billed as a failure (Taubenfeld and Jessop, 1959), an 'episode of schizophrenic colonialism' (Foster, 2008, p. 367); a 'legislative and administrative machinery [that] creaked on with comic opera pomp' (Foster, 2008, p. 370); as well as the rhyming 'pandemonium' (Grubb, 1999). Its longevity, however, may suggest otherwise. In any case, a critical evaluation of the condominium experience would 'prove instructive for those seeking to evaluate proposals for joint sovereignty over Antarctica' (O'Connell, 1970, p. 71). (One may add: and other controversial areas, like the Diaoyu/Senkakus.) From time to time, states with rival claims to territory have attempted to resolve their differences via the political expedient of joint, or shared, rule; and it can be confidently hoped that such states would continue to consider, and actively seek, a similar solution in the future.

Pheasant Island

Another curious example of a land condominium, and pertinent to the focus of this book, is the case of Pheasant island, with a land area of around 2000 m^2 (or 0.00682 km^2), also known as *Konpantzia* (in Basque), *Île de l'Hôpital, Île-des-Faisans or Île de la Conférence* (in French), and *Isla de los Faisanes* (in Spanish). This islet lies near the Atlantic Ocean end of the Franco-Spanish border, along the Bidassoa river that flows between the French city of Hendaye and the Spanish/Basque city of Fuenterrabía (Guipúzcoa province).

The islet is uninhabited and is out of bounds to visitors. A gleaming white monolith on the islet commemorates the Treaty of the Pyrenees, concluded via a signing on the islet in 1659. It fixed the Franco-Spanish border, following the mountain chain of that name as the natural boundary between the two neighbouring countries.

For centuries, the islet was a favourite place of encounter for royalty and bridal exchange. In 1615, King Louis XIII of France and King Philip IV of Spain first met their wives – the Infanta Anne and Madame Royale Elizabeth, each other's sisters – on the islet, after having married them by proxy, in Bordeaux and Burgos respectively. In 1659, King Louis XIV of France met his future wife, Maria Theresa of Spain, here; while King Charles II of Spain met his wife-to-be, Marie Louise d'Orléans, on the islet in 1679.

But what makes Pheasant Island so distinctive is not just its claim as the world's oldest surviving condominium. It is also the only known case where *sovereignty is not shared simultaneously, but alternately*. For six months a year, from August to the end of January, Pheasant Island is French. For the remaining six months, from February to the end of July, it is Spanish. This effectively means that the island has changed hands over 700 times between the two countries, without mishap.

Condominium experiences like Pheasant Island show that contiguous states can exercise joint sovereignty over a territory with which they share a common border, and all the more easily so when such a territory is uninhabited.

Hans Island/Hans Ø

Pheasant Island provides an interesting precedent and backdrop to another simmering dispute in another sea between two neighbouring countries with an unresolved border. As with the Diaoyu/Senkaku, we are dealing with a tract of land with no natural resources worth exploiting, and therefore of hardly any intrinsic value; but located in a strategic area. And we are also dealing with a situation where politicians from both sides were happy to take advantage of a low-risk, Arctic sovereignty dispute for domestic electoral purposes (Byers, 2014).

Hans Island (*Tartupaluk*, in Greenlandic); ᑕᕐᑐᐸᓗᒃ in Inuktitut; Hans Ø, in Danish) is an uninhabited rocky outcrop with a land area of 1.3 km^2 (0.5 square miles), located in the centre of the Kennedy Channel of the 35 km-wide Nares Strait which separates Ellesmere Island, Canada, from Washington Land, in northwest Greenland, a home-rule territory that forms part of the Kingdom of Denmark. Its location places Hans close to the path of potential future shipping routes, oil and gas deposits and fishing grounds, just as global warming, Arctic melt and diminishing sea ice facilitates navigation through the Arctic Ocean (e.g. Ho, 2010). The outcome of the Hans Island dispute may also influence future Arctic sovereignty disagreements: the increasingly accessible Arctic is driving nations to assert territorial claims and establish sovereignty in the region (Vuille-Kowing, 2011).

Both Canada and Denmark continue to claim Hans Island. Canada claims ownership through the British Adjacent Territories Order of 1880, which transferred British Arctic possessions to Canada (Stevenson, 2007); but the Order, unsurprisingly, does not explicitly mention Hans Island in the transfer (Kristiansen, 2013). Denmark claims ownership through Greenland: in 1933, the Permanent Court of International Justice ruled that Greenland was under Denmark's legal jurisdiction (Permanent Court of International Justice, 1933). Funnily enough, Hans Island itself is named after a Greenlander, Hans Hendrik, who worked as a hunter and assistant on several expeditions along the northwest coast of Greenland in the 1800s.

During negotiations on their maritime boundary in 1973, both Canada and Denmark claimed that the island was part of their territory. No agreement was reached between the two governments on this issue, and the matter was left in abeyance. Thus, the 2,685 km-long continental shelf treaty ratified by Greenland and Canada, then the longest shelf boundary treaty ever negotiated, sets and resolves title to the whole border, except for a stretch of 0.875 km that connects Points 122 and 123: the length of Hans Island (Denmark–Canada Treaty, 1973).

In 1984, Tom Høyem, the Danish Minister for Greenland, planted the Danish flag on the island and left a message (in Danish) saying 'Welcome to the Danish Island'. He also placed, it is said, a bottle of schnapps (brandy) near the flagpole (CBC News, 2005a). At around the same time, and unbeknown to the Danes, the Canadian company Dome Petroleum was conducting research on and around the island, looking at the area's economic potential in the form of hydrocarbon deposits, shipping routes and fishing: such activity may have precipitated this, rather friendly, disagreement between two otherwise close NATO allies. Canadian troops have also visited the island, planted the Canadian national flag, and left behind a bottle of whisky. In 2005, Bill Graham, the Canadian Minister of Defence, erected a Canadian flag, which sparked a letter of protest from the Danish government (he also left a bottle of Canadian Club, which triggered no objection). (Between each of these events, the harsh Arctic weather makes sure that most flags, and their poles, are blown away.)

Geological studies conducted by the Geological Survey of Denmark and Greenland do suggest that, geologically, Hans island has more in common with Washington Land in Greenland than with Ellesmere Island. Additionally, Denmark published a geological map showing Hans Island as geologically connected to the Greenland mainland as early as 1922. These arguments would appear to support the Danish claim (Kristiansen, 2013). On the other hand, if possession is nine tenths of the law, then one should note that Hans Island was briefly home to a Canadian scientific base during World War Two. Later, in 1953, Eric Fry, of the Topographical Survey of Canada, surveyed Hans Island and built a cairn there, in which he left a note claiming the island for Canada (Byers, 2014).

Hans Island is the only disputed piece of land left in the entire circumpolar Arctic. All other Arctic sovereignty disputes concern the maritime domain (Byers, 2014). Dubbed 'one of the world's friendliest border disputes' by National Public Radio (Reynolds, 2005), numerous websites and articles have poked fun at the mild nature of the conflict (e.g. Free Hans Island, 2015). There are '[n]o green line separating antagonists, no demilitarized zones and no patriots dashing past guard posts to plant flags in a desperate attempt to reclaim land seized by invading armies' (CBC News, 2005b). It is perhaps the light-hearted nature of the affair – celebrated in various cartoons (Dolighan Cartoons, 2005; McCullough, 2004), T-shirt designs (Zazzle, 2015) and other spoofs (Ugly Duck, 2012) – that has encouraged the search for a smart solution.

Judging from past practices, should the matter be taken to adjudication by the ICJ, the court is likely to award the island in its entirety to one party or another (Kristiansen, 2013). The most common alterative resolution, and the one circulating as most likely, is that of a split à la Chapter 7: a clear and equitable land division of the island 'down the middle' (Humphreys, 2012; Mackrael, 2012). It would be relatively forthright to draft a treaty that would simply ratify 'split ownership': dividing the small island by connecting Points 122 and 123 of the 1973 Canada and Denmark Delimitation Treaty, adding Hans to the list of uninhabited islands split between more than one country (Kristiansen, 2013); Canada and Denmark would thus share a short and very remote land border (Byers, 2014). One other option involves accepting a potential Inuit claim: if Hans Island can be considered as having never been colonised, then the Inuit might arguably retain any pre-existing, pre-colonial rights (Byers, 2014). But a third option is perhaps the most obvious, once it is seen as possible and workable: Hans Island being declared a condominium, with Canada and Denmark both having sovereignty over all of it, and with the Inuit in both Greenland and Canada involved in such a solution (Byers, 2010, p. 30; 2013, p. 15). 'It will not be a "winner takes all" situation. Because the opportunity for a win–win is apparent', argues Professor Michael Byers (in Zilio, 2012, n.p.).

Conclusion

Although often meant as expedient temporary fixes, some condominium arrangements have stood the tests of time: 'There is nothing in legal theory or in the nature of sovereignty to render impossible a permanent and agreed division of sovereignty as suggested by the very nature of a condominium' (Lauterpacht, 1970, p. 71).

Festering disputes like the status of Jerusalem, Kurdistan, the Caspian Sea, 'the Cyprus Question' – who knows, perhaps even Taiwan – might benefit from a condominium approach (Bahcheli, et al., 2004). It is disappointing that some scholars, burdened and intoxicated by the contemporary primacy of sovereignty, flatly dismiss the solution protocol of the condominium as a merely curious historical relic from a bygone age (e.g. Schneider, 1992, pp. 732–724). Additionally, problems in the administration of a condominium typically arise where human subjects are involved; issues that deal with boundary crossings, taxation regimes, citizenship and nationality. No such issues arise in the case of the Doaoyu/Senkaku which have had a population of zero for many decades now (Harris, 2012, p. 611). Finally, condominium arrangements over water present less delicate demands for joint cooperation since the states concerned are generally fretful mainly over rights and access to natural marine resources.

At the very least, the historical track record of condominium arrangements suggests that supposedly indivisible territorial rights *can* be rejigged in order to create a joint exercise of territorial sovereignty (Samuels, 2007, p. 776). They may be the outcomes of unintended consequences, or meant as merely transitional governance mechanisms until clear territorial demarcations are established; but condominia continue to represent the art of the possible and doable: 'If a condominium has been considered in the case of Antarctica, there is no reason not to consider it in the case of the [Diaoyu/Senkaku] islands' (Harris, 2012,

p. 612). Moreover, let us remember that the Ryukyu island chain, located so close to the Diaoyu/Senkaku islands and incorporating Ishigaki City, from where the contested islands are notionally administered by Japan, was incorporated into Meiji Japan as a prefecture only in 1879. It had, until then, been itself a condominium of sorts, under dual Chinese and Japanese suzerainty since 1609 (Drifte, 2014).

9 Suppressing (or suspending) sovereignty

Introduction

A natural variant of the condominium option is a 'solution protocol' where condominium features apply in spaces where conventional territorial claims have been made. However, many such spaces are *not* condominia because, rather than being territories that belong to more than one country, and to the clear exclusion of others, they effectively 'belong' to no one. We are dealing with territories that are expressions of non-sovereignty; and while various states may have lodged specific sovereignty claims to such territories or to parts thereof, all such claims remain suppressed and dormant. It may come as a surprise to readers that such a 'global commons' is the situation that prevails with regards to more than 99.999 per cent of the known universe.

International treaties and agreements are usually historically contingent. Back in the late 1950s, the main developed countries, with their unmatched technological and economic power, held the absolute advantage in entry and resource exploitation of the seabed in international waters, the Antarctic, the atmosphere and outer space. Were these four spaces to be declared *terra nullius*, then the 'first come, first served' principle would apply: the country to occupy the area first would secure sovereignty and gain exclusive rights to exploitation (Elliott, 2004, p. 146). To thwart the concentration of such rights to certain world powers, and to prevent related conflict to be kindled during the tense years of the Cold War, the world's many newly independent developing states collaborated to create a joint management system of global commons beyond and outside national jurisdiction in these four areas, under the global governance regime of the UN.

The four 'global commons'

Such efforts have helped the high seas, the continent of Antarctica, the atmosphere and outer space to be perceived as *res communis*: non-exclusive, global commons (Buck, 2012; Caldwell, 1998, p. 126). All four have been defined within international regimes as spaces outside the territory of sovereign states and beyond the normative inhabitable zones of humans. In at least three out of the four instances, their current governance regime is an outcome of the deliberations at the International Geophysical Year (IGY) of 1957–1958. It was on the basis of the experience gained during the IGY that three significant treaties emerged within a decade: the 1958 United Nations Law of the Sea Convention (LOSC); the 1959 Antarctic Treaty (later Antarctic Treaty System, ATS); and the 1967 Outer Space Treaty (OST). Each of these laid down the principles of the legal regimes for exploration and governance of these extraterritorial spaces. These innovative pieces of regulation have provided important lessons pertaining to the governance of trans-boundary systems *beyond* sovereign jurisdictions; they have also successfully calmed down political tensions over sovereignty claims – so far. But their grip may be loosening.

There are serious concerns about the wisdom and credibility of global commons. Since Garett Hardin's seminal essay on 'the tragedy of the commons' (Hardin, 1968), there have been many attempts to move away from a 'commonwealth' interpretation and management of these vast spaces. There are corporate and national state interests (these two often being intimately connected), relating to the exploitation of natural resources in what are currently frontier and extra-territorial zones (such as the high seas). There is a growing number of countries and industries, even super-rich individuals, who now have the capital, technology, will and political ambition to partake in such projects. And there is concern about the missing, effective, global stewardship over resources – such as fish stocks in international waters – allegedly held in trust for all humankind but savagely exploited in the interim. Might international law, in spite of a plethora of good intentions, itself have set up these extra-territorial spaces for ruin (Milun, 2011, p. 4)? The governance of global commons *can* be perfected (e.g. Ostrom, 1990); but the contemporary neo-liberal climate is eager to push out and extend the boundaries of territoriality. Thus, it may be just a question of time before what are global commons today become sovereign territories tomorrow.

So, for example, the US Congress passed the Space Resource Exploration and Utilisation Act in 2015 (US Congress, 2015). This bill radically revises elements of the original OST from 1967, under which any resources returned to Earth from space were considered to be the property of all humankind, and thus could not be owned, bought or sold (UN Office for Outer Space Affairs, 2015). The 2015 bill alters this so US private companies and states can maintain property rights over any resources brought down to planet Earth. It opens up outer space to venture capital and ensures that US private companies can obtain a title over any resources obtained in outer space. Mineral prospecting operations may soon begin on asteroids, the Moon and other celestial bodies.

Additionally, the Protocol on Environmental Protection (PEP, 1998) to the Antarctic Treaty can be reviewed at any Antarctic Treaty Consultation Meeting, and will be formally open to revision in 2048. There are already scores of research organisations and companies from at least 27 states undertaking such tasks as biological prospecting in the Antarctic (Salazar, 2015). Yet, who actually owns the continent's microbial diversity remains unclear and contested. With intensified national interest in mineral and biological resources, Antarctica could potentially become the world's most lucrative commodity frontier. The continent's rich resources have led to calls for equitable sharing via a declaration of Antarctica as 'common heritage of mankind' (Zang, 1991); but there are no signs of movement by the contracting parties to the ATS. The major turning point may have been reached when the 1988 Convention on the Regulation of Antarctic Mineral Resource Activities (CRAMRA) – which would have acknowledged the rights of the international community and introduced 'benefit sharing arrangements' – failed to enter into force: no country has ratified it (Jabour, 2010).

And, as far as LOSC is concerned, countries already have jurisdiction over the seabed (and its resources) beyond their 200 nautical mile (nm) EEZ when this is an 'extended continental shelf,' usually of relatively shallow water. At the time of writing, the 77 submissions to the UN Commission on the Limits of the Continental Shelf (CLCS) – that is looking at supra-200nm seabed rights – have already totalled 27 million km^2 (10.4 million square miles) and involve dozens of states (CLCS, 2016). The rules thus far have been largely redundant since the regions beyond the previous limit are so remote they have would cost too much to develop. But this is not the case any longer. Technological advances have opened up huge deep-sea possibilities: specialist firm Transocean drilled a well in a record 3,174 meters (10,411 feet) of water off India in 2013 (Chee, 2013). It was LOSC that considerably reduced the extent of the high seas as a global commons (DeLoughrey, 2007, Chapter 1).

Meanwhile, last and certainly not least, the atmosphere has soldiered on as an unregulated 'no man's land' that is openly accessible and appropriated by everyone free of charge in most regions of the world. However, the European Union and a select few other countries have started to price carbon emissions and introduce other emission trading schemes (Edenhofer, 2013). And global environmental governance has advanced at a rapid rate in this sector, focusing on ozone depletion and the Montreal Protocol on Substances that Deplete the Ozone Layer (Montreal Protocol) in the 1980s, and moving on to the institutional actions and scientific advice of the World Meteorological Organization, the Framework Convention on Climate Change and the Intergovernmental Panel on Climate Change (IPCC) (Hoffmann, 2012; Miller and Edwards, 2001).

The Moon could have easily become a fifth global commons. Inspired largely by the UNCLOS regime, the Moon Treaty (1967) shares the language and intent of its four other global commons cousins. It is an international agreement that turns jurisdiction of the Moon the other plants of our Solar System (other than Earth) and all other celestial bodies (including their orbits) over to the international community. But this Treaty has only been ratified so far by 16 states. And these do not include a *single* state that engages in self-launched manned space exploration: such as the US, some member states of the European Space Agency, Russia and India. Both China and Japan have not signed the Treaty. The Moon Treaty has to be considered a failure: space-faring states in particular object to the Treaty's requirement that extracted resources (and the technology used to that effect) must be shared with other countries.

The four extant global commons are collectively bunched together in this chapter as expressions of the (temporary) *suppression* or *suspension* of sovereignty claims. (This can also be called 'freezing': an apt term especially for Antarctica, where the temperature rarely goes above the freezing point.) These frontier sites constitute 'resource domains or areas that [currently] lie outside of the political reach of any one nation state' (UNEP, 2015). This is not to detract from their intrinsic value as elegant pieces of international regulation in their own right; and, after all, they continue to hold sway. However, they do tend to be defined by what they are not; and they each look increasingly under threat, with various countries, firms and investors chafing at the bit to extend conventional ownership and any derivative rights.

As long as they hold sway, these regimes need to be acknowledged. Together, these protocols literally cover the universe. Even if their reach is trimmed and their margins reduced, it is realistic to expect that features of the global commons philosophy will continue to hold, and over large spaces. And while they exist and operate, these regimes also challenge and debilitate the Westphalian concept of state territoriality that lies at the heart of the precepts of sovereignty (and cut it deftly down to size).

Palan (2003, p. 97) succinctly describes the process whereby a putative three-dimensional cage of state/national sovereignty has emerged. The transition makes for an interesting parallel between a distinct form of political rationality and technological advances (e.g. Braun, 2000). A state's sovereign territory originally extended laterally and horizontally into the adjoining seas, to the extreme range of a cannonball, which was (at that time) three nautical miles, and, by the early nineteenth century, extended to 12 miles (Prescott, 1975, p. 37). The same trend occurred in the vertical plane: when hot-air balloons and eventually aircraft began to roam the skies, national boundaries were extended 'up', to the von Kármán line, that is, 50.55 miles – now, 100 km (or 62 miles) (Gammeltoft-Hansen, 2014, p. 117). The principle was also applied 'down', to the entire sub-soil of national territory, and ending at the centre of the earth (Kish, 1973). Such upward, downward and lateral limits established the contours of three-dimensional sovereign space.

It is now these four final frontier spaces – the high seas, the Antarctic, the atmosphere and outer space – that are preventing further territorial creep; even as the concept itself has been found seriously deficient and powerless in the face of international migration, global terrorism, footloose capital, offshore tax shelters, global financial crisis, the dangers of nuclear fallout, environmental degradation and climate change, and the emergence of the European Union, to name some of the various policy challenges. States *do* exercise powers and influence in the territories, and amongst the citizens, of other states (Baldacchino, 2012). It is in rendering disputes like the Diaoyu/Sekaku as the abrasive grating of two, adjacent, three-dimensional sovereignty cages jockeying for power and space that opportunities for looking at realistic alternatives are stunted. The sovereignty paradigm has clear limits (pun intended).

The international treaties that govern the use and exploration of 'the four global commons' (Ruppel, 2015, p. 100), have much in common in inspiration as much as in execution. Amongst these four, however, and continuing with the deliberate island slant of this text, the focus of what follows shall be squarely on Antarctica.

The potential application of the Antarctica regime to the Diaoyu/Senkaku case will, of course, be at the forefront of this review.

Antarctica

When looking at the world's maps, with their usual projections and perspectives, it is easy to forget that a whole continental bulk lurks at the Earth's southern pole. Antarctica's land mass is around 14,000,000 km^2 (5,400,000 square miles), larger in size than both Europe and Australia. It is the continent least hospitable to human habitation. While scientific bases there are now manned full-time, the continent remains, so far, without any permanent human residents.

Antarctica holds many records: it is the coldest, most mountainous, windiest and driest continent. It is the most environmentally pristine; but also home to possibly the world's third-largest oil reserves (as many as 203 billion barrels), as well as coal, various iron ores, manganese and hydrocarbon, plus fish and 90 per cent of the world's fresh water (as frozen ice) (Winterbottom, 2014). But what concerns us most here is Antarctica's operational status as a *de facto* multi-country condominium. (I was about to write pan-dominium; but that sounds uncannily similar to pandemonium, so I changed my mind.)

Antarctica is being presented as its own solution protocol, and not alongside the other condominia arrangements that were reviewed in Chapter 8, because, from a *de jure* point of view, Antarctica is definitely *not* a condominium. It is *not* a territory that belongs to more than one country. For a territory of such a size, it is an expression of non-sovereignty: it 'belongs' to no one. Antarctica has an Internet country-code top-level domain (.aq); but it is the only land mass on this planet that is neither (part of) a country of its own right nor governed by any other one or more sovereign states. It is not sovereign, semi-sovereign or quasi-sovereign. It has no citizens. Many of its asserted 'international boundaries' are subject to multiple claims. It has no government. Legally, it would be best described as an internationalised space; a global, or international, commons (Joyner, 1992, p. 264). The continent's status is the outcome of a practical deal: it is administered by parties to the Antarctic Treaty System (ATS). The original Antarctic Treaty was signed by 12 countries and came into force on 1 December 1959; it has 52 signatories at the time of writing (August 2016), and includes both China (which signed on in 1983) and Japan (in 1961) (Secretariat of the Antarctic Treaty, 2015). The Treaty designates Antarctica as a 'natural reserve, devoted to peace and science', forbids nuclear explosions and nuclear waste disposal, supports the scientific research pursued by a multi-national community that now numbers around 5,000 personnel, and protects the continent's unique ecosystem (Conference on Antarctica, 1959).

Seen from the vantage point of today, such a turn of events is nothing short of incredible. After all, the initial scramble for Antarctica by the colonial powers was a classic sequence of land claims, mirroring the 'scramble for Africa' of the previous century. The first unilateral move was by Britain in 1923 (with the Ross Dependency transferred for administration by New Zealand; and the Australian

Antarctic Territory transferred for administration by Australia); then France in 1924; then Norway, with its whaling interests, after 1929; then Germany in 1938 (but forced to cede these claims after World War Two); and then the neighbouring South American countries, Argentina and Chile. The Argentine, British and Chilean claims all overlap in places. Apart from these eight claimant countries, Peru, Russia, South Africa and the US have reserved the right to future claims on Antarctic territory; while Brazil has designated a 'zone of interest' in the region. Interestingly, via Article 2(e) of the 1951 San Francisco Peace Treaty, Japan renounced 'all claim to any right or title to or interest in connection with any part of the Antarctic area' (Scott, 1999). Between 1943 and 1948, Argentina, Chile, UK and US all sent avowedly military expeditions to Antarctica and undertook military operations in the region (Jacobsson, 1998, p. 2).

The prevailing mythical narrative is that it was the International Geophysical Year (1957–1958) that ushered in a period of intensive scientific cooperation in the region, which laid the seeds for the consideration of a creative and unique solution to the governance of an equally unique continent. Had this narrative been true, one would have expected a more widespread pacification and collectivisation of the planet from the plethora of international programmes in earth sciences post-WWII (Bulkeley, 2010, p. 11). The idea of an internationalised Antarctica had been mooted in the mid-nineteenth century; but it was a counter-proposal by Chile that argued for an agreement on a programme of scientific cooperation that effectively suspended all sovereignty claims for its duration. This is what eventually came to pass, in spite of the persistence of international rivalries and the involvement of the then Soviet Union at the height of the Cold War.

The Japan Maritime Safety Agency was actively involved in the early burst of antipodean science: it offered the ice breaker *Sōya* as the Southern Ocean observation ship, while also building Showa Research Station as the first Japanese observation base on Antarctica: over 1,000 personnel have lived and worked there since its opening in 1957 (Otani et al., 2004). It was mainly to prevent the possibility of military conflict in the region that the US, UK, Soviet Union and nine other countries with significant interests in the region negotiated and signed the Antarctic Treaty in 1959. (Meanwhile, China only established its first Antarctic research base in 1985; but it now has four operating stations, is opening a fifth one soon, and is investing in a second ice breaker and new ice-capable planes and helicopters (Bender, 2015).) The territorial claims of the various countries have not gone away; they have been merely frozen, suppressed. Thus the Treaty keeps the existing claims in abeyance, prevents fresh claims or extensions of existing claims from being lodged, while yet acknowledging tacitly that claims exist:

> No acts or activities taking place while the present treaty is in force shall constitute a basis for asserting, supporting or denying a claim to territorial sovereignty in Antarctica, or create any rights of sovereignty in Antarctica. No new claim, or enlargement of an existing claim, to territorial sovereignty in Antarctica shall be asserted while the present Treaty is in force.
>
> (The Antarctic Treaty, Article IV)

Would the noble goals of the ATS – peace, scientific cooperation, no military activity – be protected and preserved, or undermined and threatened, with the continued suppression of these claims (Grob, 2007)?

One early challenge to the ATS dealt with environmental protection: a clear departure from the goals of the original ATS. Antarctic terrestrial biodiversity is seriously impacted by rapid climatic changes and the unprecedented expansion of the human footprint. Apart from the potential for irreversible environmental damage at the local level, these challenges are likely to act synergistically to increase the risk of the introduction and establishment of non-native or invasive species and diseases and reduce the resilience of (already fragile) native ecosystems. The ATS answer was the PEP to the Antarctic Treaty, which entered into force in 1998 (PEP, 1998). This is the main governance mechanism that regulates environmental management in Antarctica (Convey et al., 2012).

How times have changed. Antarctica's isolation is no more. Once remote, dangerous, forbidding and uninteresting, the stuff of intrepid (and often deadly) exploration, the island continent has been opened up by technology and the end of the Cold War, and subject to commercial pressures and commodification (Kriwoken et al., 2007, p. 177). New challenges, such as fishing and tourism, along with an expanded ATS membership, have transformed the politics of Antarctica. Current controversies involving illegal, unregulated and unreported fishing, the UN Law of the Sea and the creation of the 'coastal state', tourism and whaling continue to provoke serious challenges to the governance of Antarctica (Dodds, 2010). Can the 50-plus year-old ATS regime handle the new challenges effectively?

Consider tourism. Antarctica tourism has both grown and diversified over recent decades. Tourist numbers have increased from a few thousand in 1990 to more than 45,000 in 2008, and recently plateaued at around 37,000 due to the global economic recession (IAATO, 2015; Lamers et al., 2013). Formal tourism regulation has been limited because of the complex governance structure in Antarctica. Tourism management is self-regulated through the International Association of Antarctica Tour Operators (IAATO). The success of IAATO is attributed to the high degree of organisation in the sector, largely because of the perceived benefits of membership. Continued incentives for self-organisation are needed; but changing circumstances may lead tour operators to believe that IAATO membership is no longer advantageous or sufficient. Under current conditions, the Antarctic tourism self-regulatory regime is still a fairly robust institution. However, with the predictable increase in the number of tourists and operators, the institutional structure may prove weaker in the medium term (Haase et al., 2009).

Would a legal condominium solve Antarctica's growing problems? The US had proposed such a 'multiple condominium' solution in 1948, but the idea was rejected by claimant states, to whom the idea of pooling claims did not offer much appeal (Jacobsson, 1998, p. 3). There have been no public official proposals in this vein since. The main stumbling blocks here would be two. First, who gets to vote: given the limited membership of states that would exercise sovereignty, who would want to be left out? And with claimant states being mainly first-world, would not such a manoeuvre smack of colonialism? Second, the basis for decision making: would consensus voting be practical, and when would majority voting apply? One may add a third objection: the associated and irrevocable elision of Antarctica from its unique current status as non-sovereign territory.

Conclusion

The Antarctic Treaty, and its affiliated legal regime, depends on self-regulation: 'it has no teeth' (Teller, 2014). Still, this does not diminish its status as an exceptional tool designed to resolve the particular issues involved in claims to what can be considered one of the world's largest islands. The Treaty 'freezes' national claims to Antarctica: Article IV of the ATS states that parties to the Treaty 'agree to disagree' about any (including overlapping) claims, maintaining the status quo ante 1959. And the containment of critical underpinnings of nationalism – around positions of territorial sovereignty and (now defunct) Cold War polarities – may be justly credited as 'the ATS' greatest achievement' (Hemmings et al., 2014, p. 1). It does not seek to solve the extant claims to territorial sovereignty over the continent – and which are currently in abeyance – but instead crafts a global commons and serves as a tightly focused framework agreement governing the use of mineral and other resources of the continent. The limited scope of the Treaty has evolved into a system that covers most potential activities on that icy world. The Antarctica regime demonstrates that successful cooperation amongst even many states can be secured and maintained in spite of a plethora of territorial claims, some of which are conflicting and overlapping. Cooperation in Antarctica has proceeded with little outward sign of dissension, particularly from Australia, which is a claimant to 42 per cent of Antarctic territory. And yet, the system is already struggling to keep up with, let alone handle, the pressures building up both to exploit the continent's immense mineral resources, as well as to protect its fragile, and increasingly stressed, environment (Auburn, 1982; Joyner, 1987; Samuels, 2007, p. 767).

In any case, and in spite of the glaring discrepancy in land size, the parallels with the Diaoyu/Senkaku are plenty. The continental island of Antarctica has no permanent residents; several states have had minimal contact with it throughout history; and it is thought to have the potential for vast mineral resources, which current technology renders increasingly accessible and exploitable.

10 Single sovereignty but shared jurisdiction

Introduction

Other than in cases of shared sovereignty – such as condominia (see Chapter 8) – or of non-sovereignty – as in global commons (see Chapter 9), much of the world is carved up in spaces that, unless contested, fall under the purview of single sovereign states. There are, however, situations where a single state maintains sovereignty over a specific land or sea area, but then does *not* exercise complete or full jurisdiction over what goes on in that same area. This chapter reviews a number of interesting examples from recent history where such a limitation within the power and authority of a sovereign power applies, typically following the outcome of international arbitration.

We need not go far to find our first example of single sovereignty but shared jurisdiction: Okinawa, part of Japan, entered such a state of 'political limbo' and was under US administration in the immediate post-war period and only 'handed back' to Tokyo in 1972. Even after this reversion, considerable swathes of its land area remain under US effective control as military bases (e.g. Smits, 2015, p. 4).

There are various other examples of islands, and the waters around them, that have offered themselves up as spaces for such political experimentation. Here are the best known.

Single sovereignty; shared jurisdiction: five island examples

Sedudu/Kasikili Island (Botswana/Namibia)

The ownership of Sedudu/Kasikili Island in the Chobe River, in southwest Africa, has been the subject of a formal dispute between the governments of Namibia and Botswana at least since May 1996, when both governments agreed to submit their claims for sovereignty over the small island to the ICJ in The Hague. This decision was expedited by Botswana and Namibia's significant mutual interest in the prevention of poaching, an effort that had been complicated by the dispute (Schulte, 2004, p. 250). Prior to this formalisation of the quarrel, the island's 'ownership' had been disputed by colonial governments and local residents in Namibia and Botswana since the Berlin Treaty of July 1890, when the eastern boundaries of the Caprivi Strip were defined in such vague terms as 'the middle of the main channel' of the Chobe River, so as to separate the spheres of influence of Prussia/Germany and Great Britain (Salman, 2000).

An incident that precipitated the submission of the dispute to international arbitration occurred in October 1984, when a unit of the Botswana Defence Force fired on a South African Defence Force unit patrolling in a boat in the southern channel at Kasikili Island (Alexander, 1999, p. 5; Brunet-Jailly, 2015, p. 73). (Namibia would become an independent country only in 1991.)

The island in question is known as Sedudu in Botswana and Kasikili in Namibia. The island is flat, with a land area of about 3.5 km^2 and located in the Chobe River which divides around the island, flowing both to its north and to its south. The island is flooded to varying depths for between three and four months annually, following seasonal rains.

In the opinion of the ICJ, the dispute centred on the precise location of the 'main channel'. Also with claims to Sedudu/Kasikili in mind, Botswana contended that this is the channel running to the north of the island, whilst Namibia contended that the main channel is the one to the island's south. For the Court to proceed to determine which of the two channels could properly be considered to be the 'main channel', it considered the dimensions (depth and width) of the two channels, the relative volumes of water flowing within these two channels, the riverbed profile configuration and the navigability of each channel. The Court considered submissions made by both parties as well as information obtained from *in situ* surveys during different periods of seasonal flow.

The final ruling was given in favour of Botswana in December 1999, and after a relatively long period of protracted debate and intermittent threats of military action, including the formal military occupation of the island by the Botswana Defence Force. In its majoritarian ruling, the ICJ indicated that the northern channel around Sedudu/Kasikili Island would henceforth be considered as the 'main' channel of the Chobe River (ICJ, 1999a). Accordingly, the formal boundary between Namibia and Botswana would henceforth be located in the northern channel of the Chobe River. So far, this is a classic 'zero-sum' dispute resolution verdict, rewarding one party and leaving the other empty handed.

The Court, however, also agreed, this time unanimously, that seacraft from *both* countries would be allowed unimpeded navigation in *both* the northern and southern channels around Sedudu/Kasikili Island (Johnson, 2000; Shaw and Evans, 2000). To quote from the ICJ decision: 'In the two channels around Kasikili/Sedudu Island, the nationals of, and vessels flying the flags of, the Republic of Botswana and the Republic of Namibia shall enjoy equal national treatment' (ICJ, 1999b).

The Sedudu/Kasikili island tiff provides one example of a conflict between two neighbouring countries over a very small uninhabited island which, in any case, was flooded and thus submerged under water for parts of the year. Such a stand-off appeared ludicrous, if not farcical (Maluwa, 1993, p. 113). Nevertheless, the affair reached a high level of tension, preventing resolution of the problem by the disputing parties, but which catalysed their disposition to agree to jointly submit the matter to an independent third party (the ICJ) which was called in to arbitrate over the matter. The ruling, while 'zero-sum' in adjudicating sovereignty to the island, offered a 'win–win' solution to the navigation of its surrounding waters. Thus Botswana, in this case, while granted sole sovereignty over the island, was obliged to concede unencumbered navigation rights to nationals of, and vessels flying the flag of, Namibia.

A second example takes us to the English Channel, location to what are possibly the world's oldest subnational island jurisdictions, Guernsey and Jersey, and which have both enjoyed considerable autonomy since 1204 (Everard and Holt, 2004).

The Minquiers and Écréhous, Channel Islands (Christian Fleury and Henry Johnson)

The Channel Islands are composed of two Bailiwicks: Jersey and Guernsey. Each Bailiwick is a British Crown jurisdiction, originally under the authority of a Bailiff. Nowadays, while maintaining their allegiance to the Crown, the Bailiff carries out mainly ceremonial, political and legal roles. These British Crown Dependencies were ruled by Normandy from 933 until their monarch, King John (also the King of England), was driven from mainland Normandy by King Philip II (Philip Augustus) by 1204. (The nearby Chausey archipelago – southeast of the Minquiers and the closest of the islands to mainland France – was lost to the French, although changing hands during times of war on various occasions. Likewise, some of the islands within the Bailiwicks of Jersey and Guernsey have at times been occupied by the French.) The islands might be viewed as historical anomalies, being located in a pronounced indentation of the northwestern coast of France called Golfe Normand-breton (or Golfe de Saint-Malo), but their geography is particularly distinct because each Bailiwick consists of a number of islands, islets and rocks. The two largest islands of Jersey and Guernsey are nowadays highly populated and are known internationally for their financial services industries. Each of the two largest of the secondary islands of the Bailiwick of Guernsey – Alderney and Sark – has its own government and a historical and complex relationship with the main island. At the other end of the island scale, numerous rocks combine with tidal currents amongst the most powerful and wide in the world to make this marine area one of the most difficult to navigate. In this setting, the Minquiers and Écréhous reefs, whose main islets are respectively situated 20 km south and 9 km northeast of Jersey, to which they belong, are of special interest. The sea around them is particularly fast flowing and dangerous, creating marine mazes solely reserved for the most experienced navigators. Among those who visit the reefs are a handful of both French and Jersey fishers who travel there to exploit the abundant resource of crustaceans, including lobsters and spider crabs.

Even if vraicing (collecting seaweed) and quarrying have at times been significant activities in the history of both the Minquiers and Écréhous, these occupations have not been as long-lasting or as important as fishing activities. Until the middle of the twentieth century, the reefs had a more or less troublesome coexistence inbetween Jersey and France, and between Jersey and French fishers. Jersey fishers, coming from the southeastern parish of Grouville, frequented the Minquiers where they built first wooden, then stone, 'huts' (*baraques*) on the main islet of La Maîtresse. Their counterparts from the northeastern parish of St Martin did the same on several of the larger islets in the Écréhous. In both locations, the Jersey fishers were in close contact with French fishers who came mainly from the Chausey archipelago to the Minquiers and from the Carteret region in the Cotentin Peninsula to the Écréhous. This cohabitation and co-presence in both reefs on land and water inevitably led to conflicts on the ground over fishing resources, which gradually emerged into the questioning of sovereignty between France and the United Kingdom, the latter of which represents the Bailiwick of Jersey on such international territorial issues. Although the French fishers were in no particular hurry to see the opening of negotiations over sovereignty, which might potentially challenge their access to rich fishing resources, lobbying by Jersey authorities persuaded the UK to press the French government to settle the question. In view of their divergent opinions over the sovereignty of the two reefs, the two countries decided in the immediate post-war period to request arbitration from the ICJ in The Hague. In advance of the Court's decision, however, the French Republic and the UK decided in 1951 that, irrespective of the outcome of the ICJ process, the fishers of the losing party would still have a right to continue to fish at the Minquiers and Écréhous reefs. The two countries then went to court with their respective arguments. The main one from Jersey emphasised land occupation primarily by the building of huts on both reefs. Concerning the Minquiers, the French argument focused on the building of beacons on the southern edge of the reef, which was necessitated as a result of several shipping disasters, some of which resulted in heavy loss of life.

The UK won the arbitration in the ICJ in 1953, and the Minquiers and Écréhous were thereafter recognised internationally as being parts of the Bailiwick of Jersey (ICJ, 1953). This decision was a prerequisite to the establishment of an international marine border between Jersey and France, which did not occur until the year 2000. The Granville Bay Agreement, which is how it is usually referenced (UK–France Agreement, 2000a, 2000b), was then signed after lengthy negotiations (Fleury, 2006, p. 414), and the agreement on fishing in the reefs demarcated the approved fishing locations of both countries, whereas a separate agreement indicated the international boundary. The agreements offer clarification of national maritime borders on the one hand, yet on the other they provide a distinct example where customary and territorial fishing rights are blurred within the archipelagic space of two reefs, which includes land and sea. In this context, besides the exclusive and limited access zones to fishing areas, a large zone stretching from the 9.65 km (six-mile) limit surrounding the island of Jersey to the international boundary, including the Minquiers and Écréhous, was opened up to French and Jersey fishers alike.

There is no longer any question about the Minquiers and Écréhous not being a part of the Bailiwick of Jersey, at least from an inter-governmental perspective. However, the strength of the outcome of the case that was argued in the ICJ in 1953 has been counterbalanced by the actual provisions that were implemented by a Joint Advisory Committee of the Bay of Granville between Jersey and France and concerning customary fishing rights, which was established in 2004. (A Joint Management Committee was established at the same time as the Joint Advisory Committee, with the former receiving proposals from and making recommendations on behalf of the latter.) While there is an international marine border near both reefs, the actual fishing grounds within and around the reefs are part of a complex demarcation of both shared and separate fishing zones.

The Joint Advisory Committee is composed of professionals, officials and scientists of both France and Jersey, and is tasked with achieving sustainable fisheries practices in and around the reefs. It is able to block any attempt that their professional members consider as being harmful to their activity, such as a recent proposal by the Harbourmaster of the Bailiwick of Jersey to forbid all fishing activities in the immediate vicinity of the two main islets, which was rejected by the committee and later withdrawn. In contrast, Jersey and French fishers are currently working hand in hand with the Jersey authorities on the delineation of a large zone where trawling will be banned in order to protect the seabed environment.

While the Granville Bay Agreement is intended 'to ensure the conservation and effective management of the fishery resources in the area' (United Nations, 2006a, p. 102), there are other factors that contribute to the protection of the Minquiers and Écréhous. In 2005, soon after the Granville Bay Agreement was signed, both reefs became Ramsar wetlands sites of international importance. This Jersey initiative was in recognition of the bird, marine and plant life that inhabit the reefs. However, the very fishing (both professional and recreational) that is allowed to take place at the reefs from both France and Jersey has been observed by Jersey authorities as a possible threat to the reefs' unique environmental importance (States of Jersey, 2012a, 2012b).

As locations near the maritime border between France and Jersey, the Minquiers and Écréhous are a pair of reefs without a permanent population: they are within minutes of Jersey and France, yet far removed from the modern, urban and highly populated settings that have developed in Jersey and northern France. The reefs have remained a sanctuary for the bird, marine and plant life that is found in these locations, and it is such natural resources, along with an archaeological past with periods of human settlement (such as the remains of a medieval priory on the Écréhous), that have contributed to their designation as Ramsar sites. In such a context, public concern has been expressed by some environmental protection groups, heritage organisations and 'Les Ecrehous' Residents Association as a result of recent ideas put forward by the government of Jersey about the possibility of utilising the wind and tidal resources around the reefs (States of Jersey, 2011, p. 13). Being in an open channel, along with rapid and wide tidal flows, the reefs would be able to help resource the urban population of Jersey with 'green' energy, albeit near sites of environmental importance. However, such a move would be in contradistinction to the environmental protection that the reefs currently enjoy, and also possibly affecting the fishing industry, which is protected under the Granville Bay Agreement between France and Jersey.

The Minquiers and Écréhous have traditionally been utilised and explored primarily by fishers from Jersey and France. With the rise of yachting and other marine pleasure activities in recent decades, along with the expansion of the tourism industry in the Channel Islands, the reefs have seen an increase in the number of visitors from each side of the channel separating Jersey and France. More recently, adventure tourism has emerged as an attraction for locals and visitors alike who are now able to visit the reefs as part of a short excursion. With motor boats and fishing tours being made available, particularly for those visiting from Jersey, both locals and visitors can visit the reefs as kayakers or passengers in the high-speed rigid hulled inflatable boats (RIBs) that visit the reefs on most days during the warmer months of the year. Most of these latter craft travel from Jersey, with adventure excursion companies being established in 2006 and 2011. Moreover, the shallow draft RIBs are able to navigate parts of the hazardous reefs in ways usually reserved to only the most experienced mariners. Additionally, the port of Granville on the French mainland has several companies that visit both reefs on a regular basis, and uses various types of craft and offers similar experiences to their Jersey counterpart adventure tourism companies.

As sites located between Jersey and France, the Minquiers and Écréhous offer a distinct marine and land environment that is especially attractive to day trippers wishing to see a remote part of the Bailiwick of Jersey. A trip on a RIB takes just 10–15 minutes to reach either of the reefs, and the high-speed craft offer a thrill-seeking experience across the sea and moving from an urban location to one that is relatively remote. Once at either of the reefs, and depending on the height of the tide, visitors can experience extensive sandbanks, rocks, islets and marine life. There are a number of 'huts' on the main islets, and visitors are able to walk through some of the narrowest pathways between granite cottages. Both reefs occupy a vast expanse of land, sea and sand at low water, but at high water both have only the tiniest of peaks emerging from the sea and with the 'huts' almost floating on water (especially at the Écréhous).

The contestation over the sovereignty of the Minquiers and Écréhous led to the international court case of 1953, which ruled in favour of the UK. However, after many decades of various challenges to ownership, the dispute led eventually to an agreement that did recognise the reefs as British territories, but in a context that led to the establishment of special maritime zones where French access to the reefs has been allowed as a result of a later agreement based on customary rights and earlier conventions. In this context, the notion of sovereignty is one that is both challenged and extended. Sovereignty of the Minquiers and Écréhous might be recognised as that of Britain, and more specifically as part of the Bailiwick of Jersey, but it is necessarily extended because in this instance it allows shared access within parameters agreed upon by two separate jurisdictions. The Minquiers and Écréhous may be a part of the Bailiwick of Jersey, but their existence is also as a part of France, albeit not as part of French sovereign territory. With a history of enacting a number of international agreements and conventions from 1839 to 2000, France and Jersey have generally found mutual benefit over the Minquiers and Écréhous by excluding outsiders and by having a common management plan (UK–France Agreement, 2000b, p. 11). It is ironic that perhaps without a confrontation over fishing resources in the 1830s – in this case, over oysters – then the sharing of aspects of these particular reefs would never have taken place.

In this context, the Minquiers and Écréhous offer an example of islands, and in this case mostly small islets and rocks, as sites of contestation and resolution. Their island nature, along with their location as a tense trans-boundary archipelago boasting a unique marine environment, have rendered the Minquiers and Écréhous as exemplars of porous and fuzzy boundaries, now ensconced within a regimen of treaties and agreements between France and Jersey that have been struck over more than a century.

Still in Europe, but moving to the Baltic Sea, we come across the Åland Islands, an archipelago whose very particular jurisdictional character was triggered by the outcome of arbitration by the now defunct League of Nations in 1921 over neighbours but rival claimants Sweden and Finland.

Åland (Bjarne Lindström and Elisabeth Nauclér)

The Åland archipelago, consisting of some 6,500 islands, is situated in the middle of the Baltic Sea between Sweden and Finland. Åland was separated from Sweden in conjunction with the Russian conquest of Finland in 1809. Later on the archipelago was also demilitarised. At the time when Finland opted for independence (1917), the Ålanders decided to ask for reunification with Sweden because of their Swedish culture and language. The conflict was submitted to the League of Nations in 1920. For the Swedish-speakers in mainland Finland, Åland's Swedish population was an important factor in the protection of their language, and Finland was declared a bilingual state in order to accommodate the Åland solution within its borders (Act on the Autonomy of Åland, 2004; Ahlström, 2004; Åland Culture Foundation, 2010; Barros, 1968; Palonkorpi, 2015; Spiliopoulou-Åkermark, 2011).

The League of Nations decided in 1921 that the Åland Islands should come under Finnish sovereignty, but with as extensive an autonomy as possible without becoming an independent state. Guarantees were given to secure the Swedish language and culture of the islanders, now numbering almost 30,000. Åland's autonomy is thus not due to delegation or decentralization, not even derogation of power. It is rather a division of power as a product of international resolution. Sweden as a kin-state has honoured Åland's linguistic and cultural status, but never supported any further irredentist or secessionist political strivings.

The resolution in 1921 also included an international guarantee. Åland was given the right to complain to an international body. The Legislative Assembly of Åland could thus submit complaints to the Finnish government to be presented to the League of Nations. If the matter were judicial, the Council should obtain an opinion from the International Court of Justice. This mechanism was not replaced when the League of Nations collapsed, but has since been brought up at every major revision of the autonomy system by the Ålanders. Efforts continue to be made to replace the arrangement by carving out a suitable protocol in an international context.

Åland's most important international representation is the island's official presence in the Nordic Council and the Nordic Council of Ministers. The three Nordic autonomies – the Åland Islands, the Faroe Islands and Greenland – are not bound by a decision made by the Council of Ministers if the decision is within their legal competence unless they accede to the decision, which makes the organisation a pioneer in international relations (Weller and Wolff, 2005).

The Åland autonomy has a solid ground from a constitutional point of view, and the sovereignty of Finland is partly limited where international agreements are concerned. If a treaty or another international obligation binding on Finland contains a term within the competence of Åland, the Åland Parliament must give its consent in order to have it enter into force in the islands. When Finland became a member of the EU in 1995, the Åland Islands could have remained outside. However, in order to keep Åland inside the EU, Finland negotiated a special protocol safeguarding some of the most vital interests of Åland. The Preamble of the Protocol expressly provides that the derogations in favour of Åland are granted in consideration of 'the special status that the Ålanders enjoy under international law'. The wording is very general and has led to divergences of opinion between the Finnish state and Åland, including about whether the international provisions recalled by the Protocol are only those of the Åland Agreement and exercised by the holders of the right of domicile, or also those on the military status of the islands (Scarpulla, 1999).

Åland is the only territory with a constitutional position so strong that it could have remained outside the EU when Finland acceded, but decided (after a referendum on the island) to enter the Union. The EU has no *modus vivendi* for such territories with an international status: Åland has no seat in the European Parliament, and despite participating in the preparation of the national positions of Finland preceding decision-making in the EU, Åland has in fact very limited influence if its position differs from the Finnish one. Åland is, however, liable in case of Treaty violations. Åland is, according to the protocol with the EU, considered a third country outside the tax union; but, as Finland has the competence to levy the taxes, Åland constitutes no special tax area. However, duty-free shopping is still possible when travelling to or from the islands (Fellman et al., 2015).

The Åland autonomy is a more legalistic construction than most other autonomy concepts, due to several factors. Finland had been part of the Russian Empire as a Grand Duchy with vast autonomy, where the legal framework had traditionally proven to be very important. Another factor was that the Ålanders had opted to become part of another country (Sweden), and the autonomy was against the will of the Åland people (Ackrén and Lindström, 2012). The division of power therefore had to be controlled and secured from both sides' point of view.

The legislative power is divided between the Parliament in Helsinki and a Parliament formed in Mariehamn, Åland's capital, in 1922, creating two jurisdictions in Finland. The laws passed by the two legislative assemblies are hierarchically on the same level, but in different spheres. If two bodies have the responsibility for legislation, there has to be a power-sharing mechanism. The Åland autonomy is not entrenched in the Finnish Constitution, where only reference is made to the Autonomy Act

(The Constitution of Finland, 1999). The procedure for amending these two laws is the same in the Finnish Parliament.

The laws passed by the Åland Parliament are reviewed by a special body with members representing both Finland and Åland; if they do not agree on the judgment, it is referred to the Finnish Supreme Court. The final decision is made by the president of Finland who has the right to veto a law if this is not considered as falling within the legal competence of the Åland Islands.

The division of power could be viewed as a traditional division between a metropolitan state and an autonomous region. Foreign relations, foreign trade, mineral findings, units, gauges and methods of measurement, standardisation, marriage and family relations, inheritance, criminal law, judicial proceedings, the Church Code, citizenship, legislation on aliens, passports, firearms and ammunition, the armed forces and border guards, the actions of the authorities to ensure the security of the state, defence, readiness for a state of emergency, explosive substances, the issuance of paper money, foreign currencies, telecommunications, merchant shipping and shipping lanes – all these competences belong to the Finnish state.

The legislation on the flag and coat of arms of Åland, public order and security, local building and planning, tenancy and rent regulation, lease of land, the protection of nature and the environment, water law, health care and medical treatment, social welfare, farming and forestry, hunting and fishing, veterinary care, the postal service, road traffic, trade (but with several exceptions) – these competences belong to the Åland Parliament.

There are other less traditional areas, such as the police authorities. Åland is one of the few autonomous regions having its own police force, cooperating and exercising duties belonging to the judiciary system under Finnish legislation. But what are of course the most important policy areas for a minority solution based on the need for protecting the language and the culture are: education, culture, the archives, library and museum services, prehistoric relics and the protection of buildings and artefacts with cultural and historical value; as well as the right to broadcast by radio or cable in Åland.

Åland is a modern service economy with a high standard of living, low unemployment and a growing population. Its gross domestic product (GDP) is about 16 per cent higher than the Finnish national average and some 7 per cent higher than the corresponding Swedish figure (Fellman et al. 2015).

In spite of its strong economy, however, Åland deviates less favourably from most other European and Nordic autonomies regarding economic policy and the control of the tax system and its revenues. Here, the metropolitan state is granted almost complete political control. The main ingredients in the financial transfers system between Åland and Finland can thus be summarised in four points (Lindström, 2015):

- The Finnish Parliament decides on the state taxes and fees collected in Åland.
- Thc Autonomy receives a statutory share of the total state revenue.
- The part of the state taxes collected in Åland that exceeds a certain percentage of government revenues will be refunded to the Autonomy.
- The Autonomy decides over local (municipal) taxation in practice, although with some restrictions.

Legislation on municipal taxation is the only tax revenue of significance that falls under the legislative competence of Åland. Helsinki thus controls the taxation of the Ålanders and their business, which means that most of the political responsibility for the Åland economy lies outside the Autonomy's jurisdiction.

The Åland autonomy is originally secured in an international context, and thus considered robust from a constitutional point of view. From a financial and economic policy point of view, however, it is rather weak compared with most other autonomy regimes. In the Faroe Islands and Greenland, for example, there are no substantial restrictions on local taxing powers (Bartmann, 2002; Ackrén and Lindström, 2012).

The 'legal relationship' between Finland and the Åland Islands is rigidly legalistic, and thereby very different from the much more 'relaxed' relationship between Denmark and its autonomous territories of Faroes and Greenland where disputes are settled through negotiation and political discussion; and therefore also with more space for mutual understanding. In contrast, there has always been very little space for this sort of 'open' negotiation between Helsinki and Mariehamn. Furthermore, despite the international status of Åland, Finland usually strives to handle the Autonomy as a strictly internal Finnish issue. The Finnish authorities were thus originally against the official representation of Åland in both the Nordic Council and the Nordic Council of Ministers.

Thus, Åland presents a case of a strong and legally well-entrenched autonomy supported via an international arbitration award which is, however, obliged to operate with and within a fairly centralised Finnish state that does not look so favourably on Åland exceptionality.

We move now to the tip of South America, specifically to one of the world's ten inhabited divided islands, that of Tierra del Fuego. This is the site of a tense relationship between the neighbouring Spanish-speaking countries of Argentina and Chile. The dispute dragged on for over 100 years, and almost led to war in the mid-1980s.

The Picton, Lennox and Nueva Islands and the Beagle Conflict (Peter van Aert)

In 1881, decades before the vast territory of Patagonia was entirely explored and colonised by Argentina and Chile, a border was drawn that defined the division of the two nations. This border treaty was meant to perpetuate a status quo that kept the two countries from obtaining access to both oceans, while simultaneously including previous settlements within their respective national jurisdictions (Luiz and Schillat, 1998; Van Aert, 2013). Consequently, both shores of the Strait of Magellan were assigned to Chile, due to the presence of the Chilean city of Punta Arenas at its western shore. This natural canal was of international strategic relevance, being a heavily transited inter-oceanic corridor for international commerce until the opening of the Panama Canal in 1914. Therefore, the treaty declared its waters to be neutral to assure the free passage for ships of all flags (United Nations, 2006b).

Furthermore, to prevent Chile from obtaining access to the Atlantic, the Tierra del Fuego Archipelago south of the Strait was split between both nations. Article 3 of the Treaty describes the manner in which this was to be done:

> In Tierra del Fuego a line shall be drawn, which starting from the point called Holy Spirit Cape (Cabo Espíritu Santo), at parallel 52°40'S, shall be prolonged to the south along the meridian 68°34' west of Greenwich until it touches the Beagle Channel. Tierra del Fuego, divided in this manner, shall be Chilean on the western side and Argentine on the eastern. As for the islands, to the Argentine Republic shall belong Staten Island, the small islands next to it, and the other islands there may be on the Atlantic to the east of Tierra del Fuego and of the eastern coast of Patagonia; and to Chile shall belong all the islands to the south of the Beagle Channel up to Cape Horn, and those there may be to the west of Tierra del Fuego.

It was this article that created the foundations for one of the most complex and lasting disputes that sprang from the treaty and nearly led to armed aggression, known as the Beagle Conflict, which was finally resolved as recently as 1984. The treaty did not express textually, nor did it include a map to visualise, the course of the border in the Beagle Channel and its Atlantic mouth, where many islands and islets are situated, including Navarino and Picton, Lennox and Nueva (also known as the 'PLN') islands, as well as the Cape Horn Archipelago. This left room for generous interpretations by both countries, which were rapidly expressed through maps elaborated on both sides (e.g. Beagle Channel Cartography, 2016).

A century-long quarrel began with many failed intentions to reach an agreement in a context of increasing mutual tensions. At stake was not only the access to the Atlantic Ocean and the sovereignty of the continent's southernmost bits of land, but also national political stability and credibility, and economic opportunities weighed heavily on these proceedings. Ultimately, due to their geographical proximity, both Argentina and Chile's Antarctic claims became enmeshed as part of the dispute regarding sovereignty over the tip of the South American continent.

Both countries interpreted the treaty to suggest different positions of the border. Based on article 3 that constitutes the division of the Great Island of Tierra del Fuego 'until it touches the Beagle Channel', Chile argued that Argentina did not have access to any of the waters south of the island, commonly called the *dry coast principle*. Furthermore, based on the phrase 'to Chile shall belong all the islands to the south of Beagle Channel up to Cape Horn', it proclaimed sovereignty over all islands located within and south of Beagle Channel.

The *dry coast* principle rapidly lost force. Just as the Strait of Magellan was appointed to Chile due to human settlement, Argentina had a similarly strong case in favour of access to the waters of the Beagle Channel. Thomas Bridges, a British missionary and founder of the Anglican Mission in Ushuaia, Tierra del Fuego, in 1870, was the first white inhabitant of Tierra del Fuego. In 1886, two years after the opening of the sub-prefecture in Ushuaia, he became an Argentine citizen and was given a land concession from the Argentine Congress, as an acknowledgement by President Julio Roca for his work with the native people and his assistance to shipwrecks in the region of Cape Horn. This land became Harberton Ranch, which is still owned and operated by descendants of the Bridges family. It comprised a vast territory along the shores of the Beagle Channel, as well as some smaller islands in the channel, of which the largest one is Gable. Also it is said that the Bridges had asked for, and been granted, permission from the Argentine government to conduct sheep ranching on Picton Island. These antecedents were brought to the discussion on the location of the border.

Argentina also appealed to the *oceanic principle* implicit in the Border Treaty of 1856, reasoning that the doctrine of *uti possidetis* applied here as well. The argument in this case was that, since Chile never had access to Atlantic waters before its independence; the country could not possess any Atlantic coast precisely on the grounds of this doctrine. Additionally, according to Argentine research, the axis of the Beagle Channel, which is determined by drawing a line along its deepest points, leads towards the Picton Passage. This fact would mean that the eastern mouth of the channel in fact should be considered a delta that consists of two branches, and whose southern arm between Navarino and Picton Islands was the deepest. Consequently, the two islands north of this passage, Nueva and Picton, should be considered Argentine property.

On 22 July 1971, nine decades after the treaty, both nation's presidents – Allende in Chile and Lanuse in Argentina – signed an arbitration agreement that forced a binding decision on the course of the boundary, and as a result, the title to Picton, Lennox and Nueva Islands. This arbitration would be in the hands of the British Crown; but, since the United Kingdom was then involved in a geopolitical dispute with Argentina for the Malvinas (Falkland) Islands, it was decided that an independent and international court would be named, composed of five judges, appointed by the British government. Both countries were given the opportunity to argue and counter-argue in favour of their respective interests and claims over a four-year span. Finally, at the end of 1976, 14 volumes and 213 maps were presented by Chile, and 12 volumes and 195 maps by Argentina.

On 18 February 1977, the British government published the judgment of the court, which in most of the claims favoured Chile. Maritime access to Argentina ports on the Beagle Channel was guaranteed, since 'the Court considers it as amounting to an overriding general principle of law that, in absence of express provision to the contrary, an attribution of territory must *ipso facto* carry with it the waters appurtenant to the territory attributed'. The Chilean claim that all islands located entirely in the Beagle Channel belonged to Chile was declined; the border was delimited mostly along the centre of the channel, allowing both to navigate in national waters. Furthermore, both the ocean principle and the *uti possidetis* doctrine presented by the Argentine prosecutor were rejected. The court argued that the Treaty of 1881 did not define Picton, Nueva and Lennox as Atlantic islands, as Argentina claimed, but as islands south of the Beagle Channel. Consequently, they were allocated to Chile.

Chile immediately accepted the judgment published by the UK. Argentina, however, declared it 'incurably void', on the grounds of deformation of arguments, contradictions within the arguments, bias, historical and geographical errors and impartiality, among other motives. On 8 March 1978, the court clarified to both nations that the validity of the judgment could not be declined unilaterally and that such intentions 'must themselves be regarded as null, devoid of all legal force or effect. They are not capable of impairing the validity of the Award, which in consequence remains fully operative and obligatory in law' (quoted in Broches, 1987, pp. 289–290). As a consequence, the border and the islands mentioned in the judgment were in fact and by law assigned to one or either country. This left both nations deeply unsatisfied and on the brink of war.

By the time the arbitration was published in 1977, Argentina had fallen under the dictatorship of General Jorge Videla, which began in 1976. Chile had succumbed to a similar regime in 1973, with General Augusto Pinochet seizing power. A new agreement was signed in Puerto Montt, Chile, which permitted further negotiations on the enduring differences; however, the absence of mutual trust did not leave much common ground between the parties and tensions grew. Towards the end of the year, Argentina decided to prepare for war. Under the term 'Operation Sovereignty', a plan was designed to take possession of Picton, Nueva and Lennox islands and, if necessary, invade and divide continental Chile. Thousands of infantry troops, known as *Albatros*, were sent to the Argentine portion of Tierra del Fuego Island. Army boats guarded the waters and ground weapons, such as missile launchers, were installed along the Beagle Channel's coast, from the town of Ushuaia at the western border all the way to the Mitre Peninsula in the east. Today, many of these embattlements, abandoned and deteriorated over time, still can be found along the coastal landscape, and captured in some evocative photographs (Groh, 2014).

Locally, the conflict produced an awkward situation, since more than 30 per cent of the 'Argentine' population in Tierra del Fuego was of Chilean nationality. The Industrial Promotion Law, introduced in the Tierra del Fuego National Territory by the Argentine federal government in 1972, had stimulated economic activity and immigration to the eastern part of region. As a result, the numerous Chilean inhabitants were considered a threat to Argentine sovereignty. Some returned to Chile, while others stayed and ignored the tensions together with their Argentine neighbours, watching how the local towns were overwhelmed by the military presence and national publicity.

Both countries were prepared for battle when on 22 December 1978, they agreed on a new mediation, this time by the Vatican. On 8 January 1979, the Act of Montevideo was signed, in which both countries accepted to return to the status quo of 1977 and to not disturb their mutual harmony. Pope John Paul II named Cardinal Antonio Samoré as the person responsible for the mediation. In 1980, a proposal was accepted by Chile and rejected by Argentina. Espionage from both sides, national economic and geopolitical interests, and political pressure on the military regimes, prevented the threat of military action from becoming a reality. On 2 April 1982, Argentina disembarked in the Malvinas (Falkland) Islands, and re-installed its troops on the Chilean border. Chile, despite its historic support in favour of Argentina regarding its sovereignty claims to this archipelago, refrained and refused to back the motion to apply the *Inter-American Treaty of Reciprocal Assistance*, which would have allowed military support for Argentina in the Malvinas War. Only after Argentina surrendered 74 days later was it also known that Chile had accepted the UK's request to use its air space and land infrastructure.

In the midst of all these tensions, another issue arose when, at the 1982 UN Convention on the Law of the Sea, countries were allotted an Exclusive Economic Zone of 370 km (200 nautical miles) from their coastal baselines. Chile had already anticipated this international agreement in 1977, which changed the maritime map considerably: it turned all the waters between the islands into Chilean internal waters, and this left Chile with a political advantage over Argentina, in terms of economic, logistic and military potential in the region.

On 19 October 1984, the treaty's contents were published. Like previous verdicts, this treaty also recognised the legitimacy of the Border Treaty of 1881. Also, it held on to the border determined by the Arbitration of 1977. From its eastern extreme, it extended the limit by connecting six coordinates, establishing the maritime border between both countries. While the islands were recognised as Chilean, this line actually allots to Argentina most of the archipelago's Exclusive Economic Zone to the east of Cape Horn (Treaty of Peace and Friendship Map, 1984).

The treaty furthermore established that all maritime traffic would be permitted to navigate from the Strait of Magellan to the Argentine ports on the Beagle Channel, following a predetermined route, with the presence of a Chilean pilot and when previously informed to the Chilean authorities. From these Argentine ports in the direction of Cape Horn, the treaty also stipulated a course for foreign vessels, and can be taken without a Chilean pilot and without previous notice.

On 25 November 1984, the Treaty was approved by the Argentina people through a plebiscite. Four days later, it was signed in Vatican City by the ministers of foreign affairs of both nations and has been respected ever since.

The Treaty created a Bi-National Commission; a permanent organ that aims at 'strengthening economic co-operation and physical integration'. Over the years, this Commission has never altered the contents of the Treaty, while local state functionaries observe that in some points this might be desirable from a commercial or bureaucratic point of view.

In the early nineties, the three most southern provinces of Argentina, together with the two most southern regions in Chile, constituted The Committee of Southern Integration (*Comité de Integración Austral*), in order to stimulate bilateral relations of the Southern Patagonian region through regional efforts. This Committee still meets once a year, and elaborates a formal act of progress and threats facing regional integration. However, despite all institutional formalities, effective initiatives towards integration have been scarce; both terrestrial and maritime borders are still strictly guarded. Pilots of both nations still fulfil their maritime duty as described in the Treaty 30 years ago, while commercial trade between both regions remains insignificant. Although the Beagle Conflict slowly moves towards the pages of history, real integration seems an unlikely objective because of one major issue that lingers on: Antarctic sovereignty. Both nations claim a part of the huge white continent, and their claims also partly overlap (see Chapter 9). Therefore, although it formally cannot be called a conflict due to international arrangements that assure a momentary status quo, the border issue that divides Tierra del Fuego is still a latent bi-continental and bi-oceanic geopolitical struggle over Chilean and Argentine sovereignty.

The Beagle Channel has been the site of one of the most politically charged boundary disputes and – noting that the claimant states have both largely Roman Catholic populations – was finally resolved through papal mediation (Anderson, 2003, p. 170). Like the Diaoyu/Senkakus, the 'PLN' islands are small, barren and uninhabited; but, given their geo-strategic location, their title bears important consequences in terms of navigation rights, claims to contiguous islands, the delimitation of the Strait of Magellan and further to maritime boundary claims to the south of Cape Horn. Now, subject to the 1984 Treaty, Argentina has recognised the islands of Picton, Lennox and Nueva, and their appurtenant surrounding islets, as Chilean territory. However, the Treaty sharply restricts the exercise of Chile's sovereign rights: for example, the legal effects of Chile's territorial sea around these islands is limited, in their mutual relations, to a strip of 5.5 km (three nautical miles) measured from their respective base lines – rather than the customary 22.2 km (12 nautical miles) permitted in such circumstances by UNCLOS. This provision widens the area of ocean over which both Argentine and Chilean fishers can operate. The 1984 Treaty also endorses the *bioceanic principle*, thus limiting Chilean rights in Atlantic waters in favour of Argentina (Kacowicz, 1994, p. 289). There are also two reciprocal interlocking arrangements in place, whereby both Argentina and Chile have consented to have their sole sovereignty over stretches of their maritime border curtailed in order to accommodate navigation by maritime traffic belonging to other countries.

For our final detailed case in this chapter, we travel to the far European north and Svalbard. Just like the case of Åland treated earlier in this chapter, we discuss a sparsely populated archipelago that became an intriguing jurisdictional anomaly as a result of another international treaty that was proposed and accepted in the aftermath of World War One. In this case, the state that maintains sole sovereignty over Svalbard is Norway.

Svalbard (Adam Grydehøj)

The history of the Svalbard archipelago has, from the very beginning, been one of attempted, and failed, possession.

Ever since the 1596 discovery of these Arctic islands by the Dutch explorer Willem Barentsz, various countries have laid claim to Svalbard. In 1613, an English whaling operation claimed exclusive rights to the archipelago, clashing with claims of sovereignty forwarded by the kingdom of Denmark–Norway as well as the Dutch principle of *Mare Liberum*, which conveniently argued that the seas belong to everyone. Over the subsequent centuries, whalers, trappers and scientists from a range of European nations undertook seasonal activities in Svalbard despite the competing claims to sovereignty: Svalbard, located between Norway and the North Pole, was so far from any centre of cultural, economic or military power that the political negotiations taking place in European capitals (and eventually Washington, DC) were simply irrelevant.

In time, however, whale numbers dwindled, then demand for furs declined, and eventually political players were left with a conundrum: the United Kingdom, Netherlands, Prussia, Scandinavian countries, Russia and the USA all still wished to possess Svalbard, but by the second half of the nineteenth century, no one really knew what they would do with Svalbard if they ever managed to get hold of it. The Swedish–Norwegian Council of Ministers considered annexing Svalbard in 1870, but the plan was dropped when it became clear not only that Russia was opposed but also that Norway probably could not maintain a permanent settlement there.

When Norway gained independence in 1905, it argued that, although Svalbard had come to be regarded as *terra nullius*, Norway's claim held precedence since a place called Svalbard located four days' sailing from Iceland had been mentioned in a twelfth-century Icelandic text, and in the twelfth century Iceland had belonged to Norway. This recourse to toponymical politics was particularly questionable given that, at the time, Norway was alone in calling this archipelago 'Svalbard'; to everyone else, the islands were known by the (Dutch) name for the archipelago's largest island, Spitsbergen, and there is no evidence that the Old Norse *Svalbarði* referred to today's Svalbard and not East Greenland, Jan Mayen island, or some other land mass. In any case, Norway's claims to sovereignty confronted Dutch claims to special rights on account of Barentsz's discovery; Russian claims that Pomors – Russian fishers, primarily from Novgorod – had hunted in Svalbard prior to the sighting by Barentsz; and the fact that, in 1906, the Boston-based Arctic Coal Company established Longyearbyen, Svalbard's first viable permanent settlement, which – to confuse matters even further – was populated primarily by Norwegians.

Norway, Sweden and Russia proposed a joint-sovereignty solution in 1910, which was rebuffed by the Americans and Germans, and further negotiations in 1912 and 1914 likewise failed (Numminen, 2011). Svalbard's jurisdictional confusion was eventually settled by World War One. The aftermath of the war – with the Germans defeated and the Russian government unrecognised – offered the opportunity to produce an international solution that need not take into account the positions of two countries that had strongly asserted claims to the islands. At the 1919 Paris Peace Conference, the Allied Supreme Council granted Norway a peculiar kind of sovereignty over Svalbard, leading to the 1920 Spitsbergen Treaty. The treaty was originally signed by 14 states, with the Soviet Union and Germany finally signing in 1924 and 1925 respectively. The international community may in part have wished to grant Svalbard to a state that was not a major power, a state that could threaten no one. Nevertheless, the treaty was poorly received in Norway, where segments of the press and political establishment complained of the restrictions it set on Norwegian sovereignty. On the plus side, recognised Norwegian sovereignty meant that Norway could make everyone start calling the islands 'Svalbard' rather than 'the Spitsbergen archipelago'.

The key provisions of the Spitsbergen Treaty (1920) are as follows:

- Article 1: Norway has 'full and absolute sovereignty' over 'all the islands between 10° and 35° longitude East of Greenwich and between 74° and 81° latitude North'.
- Articles 2–3: Nationals of all signatory states may undertake commercial activities 'on a footing of absolute equality', subject to Norwegian legislation.
- Article 6: Those who occupied land before the treaty's signing would be granted title to this land.
- Article 7: Nationals of all signatory states may acquire, enjoy and exercise the right of ownership of property (including mineral rights) on terms of 'complete equality'.
- Article 8: 'Taxes, dues, and fees levied shall be devoted exclusively' to the administration of Svalbard.
- Article 9: Svalbard 'may never be used for warlike purposes'.

The treaty does not specify how Norway should govern the archipelago. Norway decided to administer Svalbard through the office of a centrally appointed governor.

Article 6 of the Spitsbergen Treaty allowed other states to maintain a foothold on Spitsbergen by continuing to operate existing mining settlements. Over time, ownership over these land titles consolidated as Norway and the Soviet Union purchased other countries' claims as they became available. And so, by the 1930s, only the Norwegians and Russians possessed active settlements. Today, Svalbard's only settlements are the Norwegian towns of Longyearbyen and Ny-Ålesund (with a combined population of around 2,200) and the Russian town of Barentsburg (with a population of around 500). None of Svalbard's mining towns has ever been profitable for a sustained period, and the Norwegian and Russian mining companies are both state owned and heavily subsidised: after all, mining in Svalbard has long been an excuse rather than a reason for settlement. Such an excuse is necessary because the Spitsbergen Treaty's Articles 2–3 predicate Russia's ability to be active in Svalbard upon its undertaking *economic* activities. If Russia simply kept people in Barentsburg without undertaking some kind of business, then Barentsburg's legality could be challenged. It is thus that the Norwegian administration has attempted (sometimes successfully, sometimes not) to limit the sphere of activities in Barentsburg. For example, tourist-oriented helicopter transport between Longyearbyen (the site of Svalbard's commercial airport) and Barentsburg was halted because the Norwegian courts ruled that, although tourism represents economic activity, the helicopter flights themselves do not. Norway has furthermore deployed domestic legislation to bias the ability to undertake economic activities 'on a footing of absolute equality' against Russian interests. For instance, the Svalbard Environmental Protection Act sets strict limits on Russia's ability to open new mines connected to Barentsburg; but only after Norway had already improved its own mining infrastructure (Åtland and Pedersen, 2009).

It is convenient to discuss Svalbard's population as being divided into Russian and Norwegian towns, but Russia holds no special status with respect to the Spitsbergen Treaty. Russia's government was not party to the drafting of the treaty, and Russia's involvement in Svalbard comes down solely to a Russian desire to be active there. Citizens of any state in the world have always had the legal right to enter, reside in and undertake economic activities in Svalbard; but historically most opportunities for immigration and residence were directly linked to Norwegian or Russian state interests. Although there are no national borders within Svalbard, interaction between the Norwegian and Russian settlements was formerly minimal, simply because these company towns' policies were created by their state-owned operators.

This has changed with Longyearbyen's liberalisation in the 1990s. Taking advantage of Russian political weakness, the Norwegian government sought to bolster Longyearbyen's viability by reducing the power of the state mining company and encouraging private tourism services (Arlov, 2011, pp. 324–325). As a consequence, the Governor's Office lost absolute control over activities in Longyearbyen. Businesses were established by, and employed individuals from, many countries (Longyearbyen's population of around 2,150 includes citizens of over 40 states), and the emergence of a community of permanent residents contributed to demands for local democracy; elected local government was indeed introduced in Longyearbyen in 2002 (Grydehøj et al., 2012).

Similarly, the Norwegian town of Ny-Ålesund (formerly a mining settlement) has been transformed into an international scientific research centre, hosting research stations from a variety of states that desire a foothold in the Arctic. The Norwegian government originally encouraged this process in order to disempower Russia by relatively degrading its privileged status in Svalbard; but today Norway finds that it lacks legal mechanisms to control activities by state interests from, for example, China even if it wishes to. The genie has well and truly been released from the bottle (Grydehøj, 2014).

Svalbard is a uniquely unbordered international territory and served as a negative role model for the eventual (bordered) international solution applied to Antarctica. Although Svalbard has maintained its current legal status for nearly a century, this has been courtesy of a mix of circumstance and careful geopolitical maintenance. It is easy to imagine that, had sovereignty over Svalbard been granted to a major power rather than to Norway, coexistence of Western and Soviet actors in the territory during the Cold War would have proved unsustainable. Similarly, though it is not admitted in public, the international community depends on Russia to continually test the limits of Norwegian sovereignty and to protest against Norwegian attempts to exceed their jurisdictional capacity. Indeed, it was in the aftermath of the fall of the Soviet Union (in a unique period of Russian state weakness) that Norway began pushing to enforce domestic law in Barentsburg. In issues of fisheries, exclusive economic zone, transport, continental shelf, demilitarisation, and cultural heritage, Russian steadfastness against Norwegian claims has provided a bulwark for international rights to the archipelago and has (via Norway's internationally oriented reaction) indirectly led to the true opening up of Svalbard to the international community. At present, many states are active in Svalbard, but they are only there because other states are there as well. Russia and Norway in particular have expended vast resources and political capital to maintain economically unprofitable, socially problematic and militarily useless settlements.

By now, all parties involved must know that they can never truly possess Svalbard. State activity in Svalbard thus ultimately aims to prevent others from possessing the islands. From the perspective of Longyearbyen's burgeoning and spectacularly international local community, this has brought about positive results. From an international relations perspective, Svalbard has long been and continues to be a cynical and futile exercise in statecraft as an end in itself.

The case of Svalbard thus offers interesting insights into potential strategies for resolving the dispute over the Diaoyu/Senkaku island chain. One lesson is perhaps that legal solutions cannot be future-proofed: the creators of the Spitsbergen Treaty could not have predicted the advent of large-scale Arctic tourism, much less that future scientists in Svalbard would be interacting with satellites in space to further military activities in Afghanistan – both activities that have occasioned recent clashes between Norwegian and Russian interpretations of the Spitsbergen Treaty. Any legal solution, no matter how carefully crafted, will be open to contrary interpretations and debate unless there is genuine and abiding consensus among all parties as to the laws' desired intentions and effects. If consensus existed, of course, the creation of such a legal solution would be unnecessary. Nor is a solution like the arrangements concerning Spitsbergen particularly useful even for that most fundamental of treaty goals: the avoidance of armed conflict. As with Åland, Svalbard has an enviable history of demilitarisation – except for during wartime, the time when demilitarisation is most useful. In fact, during World War Two, the Allied powers evacuated Svalbard and razed its settlements in order to pre-empt potential German occupation, and the archipelago was subsequently the scene of sporadic fighting.

Conclusion

Exasperation with conventional 'win–lose' solutions, and the intervention of third-party (often international) arbitration, have together conspired to conjure up some interesting examples of creative governance on, for and around islands. In this chapter, all cases reviewed recognise one country as exercising sole sovereignty over a specific territory; but the authority arising from such sovereign title is partial and must be shared with one other or more authorities. Tensions and disagreements over who can, or cannot, do what become the stuff of regular public policy in these jurisdictions.

11 Shelving (or setting aside) dispute while developing

> Set aside dispute and pursue joint development.
>
> (Deng Xiaoping, 25 October 1978)

> Japan has consistently maintained that there has never been any agreement with China to 'shelve' issues regarding the Senkaku Islands. This is made clear by published diplomatic records. The assertion that such an agreement exists directly contradicts China's own actions to change the status quo through force or coercion.
>
> (Ministry of Foreign Affairs, Japan, 2016b, p. 3)

Introduction

Maritime joint development zones have emerged of late as an important mechanism for overcoming deadlock, 'lose–lose' situations in relation to maritime jurisdictional claims. These episodes can be broadly divided into two classes: those concerning joint zones that have been agreed *in addition to* an agreed and accepted maritime boundary line; and those that have been crafted *in the absence of* a boundary line. The latter thus involve the joint development of resources in what are effectively disputed waters. Such an interim strategy has proved to be a popular alternative, since it allows for resource extraction and the economic benefits that it permits, while creating conditions for building confidence and mutual trust. Nevertheless, this arrangement comes with its own slate of challenges.

A number of these cooperative mechanisms predate UNCLOS: a revenue-sharing agreement from hydrocarbon deposits in a disputed maritime zone between Bahrain (an island state) and neighbouring Saudi Arabia goes back to January 1958 (Lagoni, 1979). But joint maritime zones have largely been concluded since UNCLOS was opened for signature. Such joint developments, referred to as 'provisional arrangements of a practical nature', are governed by UNCLOS Articles 74(3) and 83(3). These provide, for the states concerned,

> in a spirit of understanding and cooperation . . . [to] make every effort to enter into provisional arrangements of a practical nature and, during this transitional period, not to jeopardise or hamper the reaching of the final agreement. Such arrangements shall be without prejudice to the final delimitation.

The definition of joint zones in this manner is, nevertheless, often a sensitive and demanding affair. The exercise of sovereignty in disputed waters and/or over disputed territory can be seen to validate such claims to some extent, giving them practical import and thereby a degree of endorsement and legitimacy that they may not warrant. This condition has made some sovereign states reluctant to, in a sense, recognise and accept what are regarded as excessive claims through their practical deployment in the operations of, say, a maritime joint development zone (Schofield, 2014).

This chapter explores past experience of how this delicate issue has been dealt with, and especially where small islands are concerned. Some observations drawn from this inventory and assessment of past practice are offered, together with some preliminary considerations on the applicability of these observations to the overlapping claims of China and Japan in the East China Sea, and involving the Diaoyu/Senkaku island chain.

Shelving/setting aside dispute while developing: four island examples

Iceland–Norway (Jan Mayen Island)

In 1980, Iceland and Norway agreed on a maritime boundary that was negotiated been the parties on the basis of the recommendations of a Conciliation Commission (Agreement between Norway and Iceland, 1980). Norway assented to the Icelandic claim for an unrestricted 370 km (200 nautical mile) zone in the direction of Jan Mayen. This entailed Norway yielding almost 30,000 km^2 of sea area, which lies within 370 km of the Jan Mayen coastline (Harsson and Preiss, 2014, p. 124). In addition to the boundary line, a joint zone was also established. A particular feature of this 45,470km^2 joint zone is that it straddles the agreed Iceland-Norway boundary line in an uneven manner (Agreement regarding Iceland and Jan Mayen, 1981; Charney and Alexander, 1991, pp.1755-1765). Overall, 61 per cent of the joint zone lies on the Norwegian side and 39 per cent on the Icelandic side of the boundary line. Each state is entitled to 25 per cent of revenues deriving from the exploitation of oil and gas on the other side of boundary; while hydrocarbon fields straddling the joint zone and Icelandic waters are considered wholly Icelandic. The uneven distribution of the joint zone across the delimitation line, in favour of Iceland, took into account the huge population disparities between Iceland (over 300,000) and Jan Mayen (about 18), as well as Iceland's lack of mineral resources as compared to Norway's rich and bountiful deposits (Churchill, 1985).

Australia–Indonesia

Australia negotiated its seabed boundaries with archipelagic Indonesia in the Timor Sea in 1972, before Indonesia's 1975 occupation and subsequent annexation of the former Portuguese colony of East Timor, creating a discontinuity in the line which became commonly referred to as the 'Timor Gap'. Australia was the only sovereign state to subsequently recognise Indonesian sovereignty over East Timor: a move that can be interpreted to have improved Australia's position in claims to oil and gas deposits in the area. Indeed, Australia-Indonesia boundary negotiations for the Timor Gap were initiated, but remained inconclusive. Instead, a joint development zone – the Timor Gap Zone of Cooperation, covering an area of 60,500km², and divided into three sub-zones – was set up in 1989, and expanded in 1991, in large part based on the overlapping claims of the parties (Treaty between Australia and Indonesia, 1989; Charney and Alexander, 1991, pp. 1245–1328). While it lasted, this agreement was widely touted as 'the most sophisticated and comprehensive maritime joint development zone in the world' (Schofield, 2014, p. 88). It was superseded with the establishment and recognition of East Timor as a sovereign state (see below).

Australia–East Timor (Timor Leste)

A new agreement, the Timor Sea Treaty (TST), was signed between Australia and East Timor on the very same day that East Timor became an independent sovereign state, on May 20, 2002. This was no coincidence: revenue from seabed exploitation was to (hopefully) largely finance the new country's independence. The agreement established a joint authority model, the Timor Sea Designated Authority (TSDA), responsible for the administration of all petroleum-related activities in the disputed area of the Timor Sea. The TST also established a Joint Petroleum Development Area (JPDA), which coincides with the central sub-zone of the former Australia-Indonesia joint development zone. In the past, revenues from this 'sovereignty neutral' zone had been shared on a 50-50 basis between Australia and Indonesia; but, under the TST, revenues from seabed resources exploited within the JPDA are split 90-10 in East Timor's favour.

So far, so good. But complications arose, especially in relation to the Greater Sunrise complex of fields along the JPDA's north-eastern limit (Schofield, 2005). A revised treaty was required to establish a further interim resource-sharing deal in 2006, providing for the equal sharing of revenues deriving from the upstream exploitation of petroleum resources within the contested zone (Treaty between Australia and Timor Leste, 2006). The 2006 Treaty is without prejudice to either side's claims to maritime delimitation, and includes stringent requirements for a moratorium of up to 50 years on claims while the treaty remains in force.

Relations between Canberra and Dili took a nose dive, however, in 2013, in the wake of the Australia-East Timor spying scandal. East Timor launched a case at the Permanent Court of Arbitration in The Hague to pull out of the gas treaty it had signed with Australia, accusing the latter of having its intelligence agency bug the East Timorese cabinet room in Dili in 2004, and therefore having unfair advantage in negotiations (ABC, 2013; Cleary, 2007). Negotiations about the maritime boundary are ongoing (Gullett, 2016): East Timor has been seeking to have the treaty declared invalid in order to establish a permanent maritime boundary halfway between the two countries, which would put more of the Greater Sunrise oil and gas field within East Timor's EEZ (Coutinho & Briosa e Gala, 2014; Everingham, 2015). In any case, the good faith that justified the pursuit of cooperation for mutual benefit by the two countries has been severely compromised.

South China Sea?

In areas with multiple claimants, such as the South China Sea imbroglio, it may now be unrealistic for anyone to expect the parties involved to resolve the complex disputes through negotiation, especially if sovereignty disputes are also implicated. Moreover, sovereignty disputes cannot be subjected to third party dispute settlement without the consent of the disputing parties; and Beijing has made its position on this matter quite clear. Given the acute national sensitivities associated with sovereignty disputes in the South China Sea, it is generally agreed that the most viable interim solution for managing the disputes there is for the claimants to set aside and shelve their sovereignty disputes and focus on less contentious issues (Zou, 2006; Simon, 2012). Such initiatives would include cooperation on marine biodiversity and the safety of navigation. Through this non-confrontational, non-binding and incremental approach, trust and confidence may yet be nurtured among the claimants and this may get them closer to consider additional cooperative measures (Beckman et al., 2013).

It therefore comes as no surprise to note that an approach that involves the joint exploitation of resources while shelving the sovereignty question appears to be a promising way of handling the Diaoyu/Senkaku affair. Indeed, this has seemed to be the case all along with the China-Japan island dispute, until Tokyo started arguing that there is *no* official dispute and the implicit acceptance of shelving the matter, as suggested by Deng Xiaoping in 1978, was a unilateral Chinese proposal that was never accepted or approved by Japan.

The Diaoyu/Senkaku territorial dispute shelved – or is it?

One may argue that what had kept the Diaoyu/Senkaku territorial dispute between Tokyo and Beijing under relative control from 1972 and for various decades thereafter was a tacit understanding (in Japanese, *anmoku no ryokai*) in both 1972 and again in 1978 to 'shelve' the dispute (*tana age* in Japanese; *gezhi* in Chinese). Such an agreement to put the dispute aside may have helped to contain the territorial conflict for so long, and in spite of several incidents that risked revitalising it. However, the Japanese government later explicitly denied such an understanding. And yet, some other, similar 'understanding' may have been found in order to climb down from the very dangerous level of confrontation that existed in late 2012, and forced the world to hold its breath.

Even though both Taipei and Beijing registered their sovereign claims to the Diaoyu Islands in 1971, this did not prevent Beijing and Tokyo from normalising diplomatic relations the following year; moving on to secure a comprehensive Peace and Friendship Treaty in 1978. On both occasions (1972, 1978), it was the Japanese side that raised the issue of the Senkaku islands, and agreed to proceed nevertheless to a conclusion of the respective negotiations despite diametrically opposed claims to the ownership of the disputed island chain. In other words, it seems that both governments agreed to shelve the issue, albeit not in writing or in any public or legal form. In the case of the September 1972 negotiations, the then head of the China Division in the Ministry of Foreign Affairs in Tokyo, Hashimoto Hiroshi, stated that when Japanese Prime Minister Tanaka Kakuei brought up the subject, Chinese Prime Minister Zhou Enlai's reasoning was that it would not be opportune to bring up and discuss the issue of the Diaoyu/Senkaku Islands, saying 'Let's discuss this some other time' (e.g. Asahi Shimbun, 2012). With the normalisation of diplomatic relations between the two countries as the paramount objective, it made sense for both sides to agree to keep the Diaoyu/Senkaku issue firmly under wraps, without having conceded anything to the other side.

When both sides negotiated the Treaty of Peace and Friendship in 1978, there appears to have been a similar willingness to put the island matter aside in order to achieve the conclusion of the Treaty. Here, we have evidence of a longer statement by PRC Vice Premier Deng Xaoping:

> It's not that China and Japan do not have any problems. For example [there are] the Diaoyu Islands and continental shelf issues. Don't drag them in now: they can be set aside to be calmly discussed later and we can slowly reach a way that both sides can accept. If our generation cannot find a way, the next generation or the one after that will find a way.
>
> (Cited in Fravel, 2012, n.p.)

The most common conclusion from these and other accounts is that China and Japan did agree to shelve the territorial issue, while in no way abandoning their respective claims to the islands, nor accepting each other's position and claims. Otherwise, the normalisation of diplomatic relations in 1972 and a Treaty of Peace and Friendship in 1978 would have been seriously compromised. It is also plausible to assume that both sides knew that there *was* a territorial problem; otherwise there would have been nothing to 'shelve'. Moreover, any such implicit understanding to 'shelve' the issue worked to Japan's advantage because it maintained the status quo, whereby Tokyo effectively administered the islands. Even so, the Japanese government did not allow prospecting and drilling for oil or gas reserves around the islands, effectively limiting the exercise of its imputed control. Nevertheless, Tokyo could not stop every single incursion on the islands, and the 2012 nationalisation of the three islands in the Diaoyu/Senkaku group – while meant to prevent even worse consequences for China–Japan relations, were the islands to fall into someone else's hands – was an implicit act of sovereignty that was seen in Beijing as the last straw in a sequence of such slights. We know what happened thereafter.

Shelving, however, need not exclude joint exploration for mineral and other resources. As long as the two sides 'agree to disagree', they can still secure a framework agreement that would allow them to jointly exploit the allegedly lucrative oil and gas resources to be found in the waters surrounding the Diaoyu/

Senkakus. China had begun exploration activities in the East China Sea, discovering the Pinghu oil and gas field in 1983, and starting extraction in 1998, some 70 km on the Chinese side of the median line. And in 1995, Chinese companies discovered a significant group of oil and gas fields in the Okinawa trough, with Chunxiao/Shirabaka being the largest gas field in this group. Oil and gas resources may lie across the median line under the seabed; this generated Japanese concerns about losing its resources on the Japanese side of the line, and led to bilateral consultations on the issue in October 2004. Tokyo also issued a licence to Teikoku Oil, a Japanese oil firm, in July 2005 to drill a test well in areas that are on the Japanese side of the median line. The outcome of these investigations suggested that the Chunxiao/Shirabaka gas field did indeed straddle the median line. But not much more could follow since Beijing protested strongly about what it saw as incursions into its own sovereign space. (Note that Japan contends that the boundary, both for the continental shelf and EEZ, should be established on the basis of the median line to be drawn between the baselines of both states, while at the same time claiming that it is entitled to a 200-mile continental shelf. China claims that the Chinese continental shelf extends beyond the 200-mile area to the Okinawa Trough, northwest of the Ryukyu Islands, and that a delimitation based on the median line is not acceptable.)

Proposals for joint development of oil and gas reserves in the East China Sea had already been mooted in the early 1970s, soon after the existence of such reserves had been announced. In 1970, Japan had reached an agreement in principle with South Korea and Taiwan (still recognised at the time as the legitimate government of China by Tokyo) to set up a joint development project. The agreement with South Korea did indeed come into force (in spite of Beijing's objections) in 1978 and demarcated a joint development zone. Seven explorations on three sites were undertaken here during the period 1980–1986, but no commercially viable deposits were found (Schoenbaum, 2008, p. 14). In 1996, PRC Foreign Minister Qian Qichen suggested (again) that the sovereignty issue over the islands be shelved and joint development projects considered (ibid.).

And what about Tokyo? During bilateral consultations in September–October 2005, Japan proposed to China a joint project of exploiting natural gas resources in an area that would however include the Chunxiao/Shirakaba wells. China rejected this and counter-proposed in March 2006 for a joint exploitation project in two other areas, one of which was close to the Diaoyu/Senkaku group. It was Tokyo's turn to decline. Japan continues to argue that that it would only consider joint development efforts with Beijing if and when the maritime border between the two countries is settled, and not before (Guo, 2010, p. 8).

Japan's current dilemma is that its unilateral development of hydrocarbon resources from the disputed areas is technologically not feasible; while it cannot effectively prevent China from exploiting the oil/gas on the western side of the median line (Guo, 2010). Thus, it may be in Japan's economic interest to push for some kind of joint development. At the same time, as long as the market prices of oil and gas are driven down – near 'six year lows' at the time of writing (Friedman, 2015) – there is certainly no financial incentive in embarking on expensive and risky extractions from the East China Sea, and particularly from the deep Okinawa Trough, for both China and Japan.

Conclusion

Japan does have one other 'island dispute' that is officially shelved: this concerns Ieodo, formerly known as Socotra Rock, a submerged reef that has been built over by Korea to house a research station and helipad (see Chapter 3). No part of the rock is visible above sea level so this is not officially an 'island'. Formerly recognised to be in international waters, the rock became an issue in 1996 when the international community extended the maximum reach of EEZs to 200 nautical miles (370 km). This created overlapping zones among China, Japan and Korea in the East China Sea. National maritime boundary delineations have made no headway in at least 16 rounds of negotiations (Kim, 2012). But the 1974 Korea–Japan Joint Development Agreement set aside the Gordian knot of sovereignty determination. Article 28 makes this quite clear:

> Nothing in the Agreement shall be regarded as determining the question of sovereign rights over all or any portion of the Joint Development Zone or as prejudicing the positions of the respective Parties with respect to the delimitation of the continental shelf.
>
> (Agreement between Japan and Korea, 1974; see also Miyoshi 1999, p. 12)

Meanwhile, any joint development by China and Japan on and around the Diaoyu/Senkaku islands could contemplate activities that have nothing to do with oil and gas extraction. There is already a five-point fishing agreement in place since 2013 negotiated between Taiwan and Japan (and which rattles China). Meanwhile, the extraordinarily rich fauna in the region, including some endangered species on the disputed islands themselves, may offer some interesting and doable possibilities connected to trans-boundary tourism: more about this in Chapter 14 and the double epilogue.

12 Swapping and selling

Introduction

When, in 2004, China and Russia were putting together the finishing touches to the division of the islands of Bolshoi Ussuriiskii/Heixiazi (see Chapter 7), there were still negotiations going on in secret, exploring other attempts at reaching a mutually acceptable solution without, however, resorting to a partitioning of that island. Such efforts failed, but not before both sides offered some interesting alternatives. Russia, for example, proposed providing China with some surplus destroyers in exchange for the whole of Heixiazi Island (Iwashita, 2005b). China refused, and in turn suggested that it was prepared to give up the whole of Heixiazi island in exchange for some territory around the Tumen River, which would provide China with access to the Sea of Japan (and an access to which China remains cheated). In this particular opportunity of give-and-take, it was Russia's turn to decline the offer (Hyer, 2015, p. 148).

Small islands are not just critical locations for military strategists, or symbolic residues of nationalism. They can also be considered as land ripe for development, sale or exchange. Upon visiting the small Caribbean island of state of Grenada, former US Secretary of State George Schultz exclaimed that the island was 'a lovely piece of real estate' (Taubman, 1984). When seen in this light, islands – whole islands – become pawns that can be sold or bartered, negotiated in cash or in kind, to secure other desirable, positional goods. Such transactions are even easier to contemplate and actuate when such islands are not populated.

This chapter examines the manner in which islands can and have been sold or bartered in this way. The bartering can also include other islands, so in such cases we deal with the swapping of one or more islands with alternative pieces of real estate which could also be islands. These 'transactions' are especially viable when there are no island populations involved in the swap. However, the most dramatic case of all involves two island units that each had a resident population which was summarily 'betrayed' in the context of superpower negotiation in the late nineteenth century. Islanders on two continents, far apart, both found themselves swapped, ceded to another power, in a clinical exchange that was handled almost like a 'book balancing', accounting exercise. More about this below.

Tradable assets

There is nothing new about islands being traded as assets. In 1763, at the Treaty of Paris, France swapped and took back the Caribbean islands of Guadeloupe and Martinique, which had been captured by the British during the Seven Years' War, for almost all of what was left of continental French North America: the only lingering residue of what had been a huge French empire in North America are the two islets of St Pierre et Miquelon (Calvert, 1994, p. 8). Still in the Caribbean, the now sovereign island state of St Lucia changed hands 14 times between the British and the French, before finally being ceded to Britain in 1814 (Anthony, 1998). Perhaps Tobago holds the absolute record. Over the years, the Dutch, English, Spanish, Swedish and French considered Tobago a tradable asset: the island changed hands 33 times between 1607 and 1800, sometimes peacefully and sometimes via armed conflict, before the Treaty of Paris ceded it to Britain, also in 1814 (DeLamater, 2011, p. 8).

Selling islands: whether real or artificial

As for the outright sale of islands, the twenty-first century has seen the emergence of a whole industry of islands, natural or artificial, being bought (and sold) by wealthy individuals keen on investing in a prime yet secluded piece of paradise beyond the sights of pesky paparazzi. Necker island is a 74-acre (300,000 m^2) property in the British Virgin Islands that is entirely owned by entrepreneur Sir Richard Branson. Actor Johnny Depp reportedly paid US$3.6 million to purchase Little Halls Pond Cay, Bahamas, in 2004. Close by, Nicholas Cage is said to have paid US$3 million for 40-acre Leaf Cay, 'an undeveloped paradise of long sandy beaches, tropical palms and even a little freshwater pond', in 2006 (Schlicher, 2012); it has been again for sale, with an asking price tag of US$7 million. Marlon Brando owned the island of Tetiaroa, land area of 6 km^2 (2.3 square miles), in French Polynesia. Leonardo DiCaprio bought Blackadore Cay, Belize, a 3.5 km (2.2 mile) long island for US$1.75 million in 2005. Mel Gibson purchased one of the world's largest private islands: 5,400-acre Mago island, in Fiji, in 2005, for US$15 million. The Robinson Crusoe character is typically male, in real life as much as in fiction (Weaver Hightower, 2007). But females can own islands too. Singer Shakira is the owner of 700-acre Bonds Cay, in the Bahamas, which cost her US$16 million; while Canadian singer Celine Dion had a mansion on her own Île Gagnon, in the Canadian province of Quebec – she sold both island and mansion for a cool Can$25 million in 2016 (*Montreal Gazette*, 2016).

'The World', Dubai (Pamila Gupta)

For an extreme extrapolation of this contemporary craze, look no further than 'The World', in Dubai. Conceptualised under the direction of Dubai's Crown Prince Sheikh Mohammed and unveiled as a tourist development in

2003, the 'World Islands' development takes its cue from the notion of (visual and environmental) excess and gigantism that characterise Dubai itself. The project reorients our thinking on the nature and culture of artificial islands and archipelagos, produces new beach-front prime property out of dirt and sand, and introduces processes of miniaturisation and faking in relation to island tourism. However, this island map *of* the world cannot exist *in* the world without environmental, ecological and economic costs. While Dubai is home to several artificial island projects, including the (also human-made) Palm Islands and the Dubai waterfront, the distinctive World Islands Development is its most ambitious project to date. 'The World', as it is called by developers and tourists alike, consists of an artificial archipelago of various small islands constructed in the rough shape of a map of the five continents of the world, located 4 km (2.5 miles) off the coast of Dubai. The islands are materially composed mainly of sand dredged from Dubai's shallow coastal waters. They range from 14,000 to 42,000 m^2 in area, separated from each other by a mean distance of 100 metres. The entire 'World' covers a land area of 54 km^2 (21 square miles) and is surrounded by oval-shaped breakwater islands. Roughly 232 km (144 miles) of shoreline was created in the process of building this tourist haven, and development costs were estimated at US$14 billion back in 2005. This completely human-made fantasy archipelago was completed in January 2008: within a few months, some 60 per cent of its islands had been sold to investors, a development that was greatly helped by Sheik Mohammed's earlier 'freehold revolution' decree of 2002 which allowed foreigners to buy luxury properties outright, and not just on a 99-year lease. This massive real estate boom lasted until late 2008, when the effects of the global financial crisis finally started to bite. 'The World' project was then suspended briefly until the Dubai government provided US$8 billion in additional financing, enabling the developers to complete their tourist-futuristic vision of 300 small private islands divided into four categories: private homes, estate homes, dream resorts and community islands, all exclusively built for the 'more well-heeled expatriates' of the once and forever 'gilded cage' of Dubai (Ali, 2010, p. 8). Formula 1 driver Michael Schumacher is the owner of one such artificial island: it was given to him as a gift by the Dubai Crown Prince and is said to be worth US$7 million. That each independently owned island (and thus each owner) is part of envisioning a miniaturised archipelago of the world contributes to The World's tourist appeal as an 'imagined community' (Anderson, 2006). In other words, 'The World' invites discourses of the miniature; '[it] celebrates new technologies at the same time that it reduces the world to microcosm but continually refers to the physical world itself, and infinite time' (Stewart, 1993, pp. 39–44).

The idea of constructing and owning artificial islands is doubly 'new': both with regard to the 'scale of their self-proclaimed spectacular appeal' as well as the 'new assemblages of exclusion and inclusion' that result (Jackson and della Dora, 2011, p. 297). Long useful as prisons and quarantine stations to house the socially suspect and undesirable, islands have discovered a new lease of life of late as 'spectacular enclosures of hope' (Jackson and della Dora, 2011) for those who can afford them, and for whom the undesirables are, in a perverse turnaround, outside and not inside (Cameron, 2012).

The consideration of islands as tradable assets is easier to condone when such islands are uninhabited and have no local population. It is a different matter altogether when political deals are struck that involve inhabited islands, and over which the affected islanders may have no say.

In the twentieth century, complete island populations have been forcibly removed in the name of science and military strategy. One such case is that of the Ilois (or Chagossians) evicted from the Chagos archipelago, British Indian Ocean Territory, in the 1970s to make way for the construction of a formidable US military base at Diego Garcia (Harris, 2015; Vine, 2011). Another example hails from the Pacific, where the residents of Bikini and Eniwetak atolls, Marshall Islands, were removed in the name of research and nuclear arms testing (DeLoughrey, 2013); Bikini and neighbouring Rongelap remain uninhabitable (*The Economist*, 2013).

Island transactions in the Caribbean

Land transactions that transfer sovereignty were once more common and remain perfectly legal under international law. The Louisiana Purchase by the US from France in 1803, and the Alaska Purchase again by the US from Russia in 1867, are amongst the most notable sales. In 1916, Denmark handed over its complete West Indies territories – comprising the islands of St Croix, St John and St Thomas – to the US for a price tag of US$25m (Tansill, 1968). This island trio comprises the modern-day US Virgin Islands.

Islands of Abu Musa and the Greater and Lesser Tunbs (Iran vs United Arab Emirates)

Ever since the time of the Shah, Iran and the United Arab Emirates (UAE) have been locked in dispute over the ownership of three strategically important islands near key tanker routes to the Strait of Hormuz: Abu Musa (population: around 600), Greater Tunb Island (or Tunb At Kubra; population: around 350) and Lesser Tunb Island (or Tunb At Sughra; population: none). The islands were first occupied by the Shah after British withdrawal from the Gulf in the early 1970s, and were then occupied by Iranian troops in 1992. In 1995, the Iranian Foreign Ministry claimed that the islands were 'an inseparable part of Iran' (Cordesman, 2007, p. 3).

Iran has hardened its stand on its sovereignty claims over the three islands of Abu Musa and the Tunbs for three reasons: oil, security and internal politics. Transportation of 20 per cent of the world oil takes place through the Strait of Hormuz, where these small islands are located. Iran has deployed heavy armaments on the islands, including Scud missiles and Russian Kilo-class submarines. The Gulf Cooperation Council and the Arab League support the claims of the UAE over the islands, while Saudi Arabia wants to refer the issue for adjudication to the International Court of Justice. Meanwhile, Teheran adamantly refuses to accept arbitration or mediation (Al-Alkim, 2002; Caldwell, 1996). However, in March 1998, press reports suggested that Tehran has offered the Lesser Tunbs and a joint administration of Abu Musa to Abu Dhabi, in exchange for the UAE's acceptance of Iran's sovereignty over the Greater Tunbs (Press Review, 1998, quoted in Foley, 1999, pp. 31–32).

This dispute is best reviewed in its historical context, with nineteenth-century Britain keen on undermining Iran – the regional hegemon – and its tight grip over the islands in the Persian Gulf and instead claiming them for its erstwhile Arab protégés.

Swapping Zanzibar for Heligoland

Probably the most dramatic and straight island-for-island 'swap' ever to take place only materialised a few months before it actually happened.

Late-nineteenth-century imperial Britain was concerned with Germany's colonial expansionism in Sub-Saharan Africa (Gillard, 1960; Langer, 1951). The eventual Anglo-German agreement of 1890, known as the Treaty of Heligoland–Zanzibar, resolved the numerous and often tortuous colonial issues arising from the imperial ambitions of both countries in the African continent. In the course of ongoing negotiations, Britain offered the cession of the island of Heligoland, with a land area of 1.7 km^2 (0.7 square miles) in the North Sea, as an inducement for the Germans to relinquish their claims to the archipelagic Sultanate of Zanzibar, with a land area of 2,650 km^2 (1,020 square miles) along the east coast of Africa. Neither the then 2,000 or so inhabitants of Heligoland nor the close to 200,000 inhabitants of Zanzibar were informed about, let alone consulted or involved in, this significant decision. The islands were 'capriciously sacrificed like pawns in an arrogant diplomatic chess game' (Drower, 2002, p. x).

For the British, the island of Heligoland had, by 1890, become a financial burden, offering no prospects of an economic return and little to the defence of Great Britain. Parliament had rejected the Admiralty's assessment of Heligoland's value and refused to vote sufficient funds to build a harbour or to fortify the island. The Germans, in contrast, placed a far higher value on the island; they had made several earlier attempts to reclaim Heligoland, and their desire to acquire the island, for defensive purposes as much as for nationalistic ones, was noted in various diplomatic exchanges. The song that served as the old German national anthem – *Deutschland, Deutschland, über alles* – was written on Heligoland:

ironically, in 1841, while the island was still in British hands. Thus, its cession heralded the end of Anglo-German rivalries in Africa (Drower, 2002, p. 44).

The British had taken over Heligoland in 1807, during the Napoleonic Wars, and this was confirmed by treaty in 1814. These same wars, however, created a spirit of nationalism that developed into a movement for German unification. As a sense of German nationhood spread, demands for the uniting of Heligoland with the Fatherland increased. Thus, to the British, Heligoland was, under the right conditions, expendable; whereas, to the Germans, it was a prize worth pursuing.

Thousands of kilometres away stood Zanzibar, an archipelago of various small islands and two larger ones: Unguja (the main island, referred to informally as Zanzibar; the birthplace of Farrokh Bulsara, better known as lead Queen singer Freddie Mercury) and Pemba. This island cluster lay off what had become in the 1880s German East Africa, part of the recent German foray into colonialism, and what would later be known as Tanganyika. The archipelago was noted for its prized cloves and other spices, as well as for its ivory trade. German efforts to suppress slaving there were resented by Arab traders, resulting in armed clashes, and leading also to significant anti-German resentment on Zanzibar itself. With the Zanzibar–Heligoland Treaty, Zanzibar was declared a British protectorate and the slave trade was banned, though ownership of slaves remained legal for some more years. What needs to be made clear here is that Germany never exercised a protectorate over Zanzibar; in 1890, it only accepted the establishment of a British protectorate over that hitherto independent island sultanate.

The Zanzibar–Heligoland Treaty was a comprehensive settlement that recognised British and German interests and spheres of influence, and established colonial boundaries in Africa: it managed to keep the peace between the competing colonial powers in the region until 1914. Without the inclusion of the Helgoland–Zanzibar 'swap' (as it continues to be known), such a treaty, most likely, would never have been agreed to (Yokell, 2010).

Conclusion

The acts of selling and/or swapping do not just suggest solutions to island disputes; they may also exacerbate the very same disputes, in spite of noble intentions. The 2012 spike in tension between Beijing and Tokyo over the Diaoyu/Senkaku islands occurred after the Japanese government nationalised the three islands in the group that were still notionally in private hands. The price tag was US$26 million. 'This is just the ownership of land, which is part of Japan's territory, moving from one [private] owner to the state, and should not cause any problem with other countries', reported Chief Cabinet Secretary Osamu Fujimura (BBC News, 2012b). It *did* cause problems, however, since the acts of sale and purchase were implicit acts of sovereignty by Japan, to which China took very strong umbrage. One wonders whether a different kind of sale, or swap – or even a part sale or part swap, involving *some* of the islands in the cluster – might offer a solution to the Diaoyu/Senkaku dispute?

13 Status quo (or stalemate or standstill)

> On strategic grounds, if a nation is willing to back down on an issue related to one's proclaimed sovereign territory, that nation will lose face, thus undermining its positions on other matters. So, in any given dispute, a country like China or Japan, with multiple island disputes, has to worry about the implications for future negotiations not only with the rival disputant but with third party onlookers.
>
> (Christensen, 2015a, p. 106)

Introduction

The politics of confrontation is a dangerous game that can be played in many ways. While parties may register a dispute over who owns a particular territory, they may claim publically to wish to reach a sensible and acceptable solution. However, the non-resolution of a dispute, and its continued festering status, may provide valuable ammunition to both parties.

One could argue that, if the Diaoyu/Senkaku dispute is actually a problem to and for China and Japan, then the solution lies in these interested parties approaching a situation, or mind set, of mutual benefit. This is not rocket science: wiggle room for negotiating and securing deals that provide a 'win–win' solution with mutual benefits exist, as this text has hopefully shown. The pursuit of common development has been mentioned time and time again. And yet, in spite of much sabre rattling and ominously dramatic moves and noises featured in the international media, international disputes of this sort may be relatively harmless. In other words, the disputes may not even be festering – suggesting the need for action, as the title of this book implies – but simply side-lined, put aside, but consciously and possibly deliberately left unresolved. As long as they do not get out of hand, high-profile international disputes, such as the Diaoyu/Senkaku case, offer the states involved an opportunity to posture and project authority without the concomitant risk of actually ending up in an armed conflict. Since the road to hell is paved with good intentions, it may prove better to err on the side of cautious inaction: so much better to live in and with an imperfect present while claiming unambiguous sovereignty (for all its worth, when another party to the dispute is doing exactly the same) than to possibly risk it all for the off chance of genuine success.

It is thus perhaps that Canada and Denmark are content to have made symbolic gestures towards occupying tiny Hans Island (see Chapter 8); that South Korea makes the most of its impractical possession of the tiny disputed islands of Dokdo/Takeshima; and that Russia and Norway continue to spar over their respective powers in Svalbard (see Chapter 10), despite neither side having any compelling reason to be there.

Stalemate as working solution

Scholars may have a proclivity for proposing neat principles as the basis for tidying things up, and wrapping up loose ends. I am fully aware that this book, and its optimistically sounding 'solution protocols', may be seen in this light. But *realpolitik* is different. As with the Falklands/Malvinas, in theory, both countries could agree to a variety of formulae; in practice, probably not (Tenembaum, 2012). Disputes can remain unresolved for many decades, and not just because of a slow and foot-dragging arbitration or judiciary mechanism: a lawsuit between the Free Imperial City of Frankfurt (in modern-day Germany) and its neighbour, the County of Hanau, over rights to a wine tithe lasted some 212 years (Seidl-Hohenveldern, 1988). Of all the solution protocols explored in this book, the stalemate situation that prevails at present in the East China Sea cannot be disregarded and excluded from serious consideration. The current status quo, fragile though it may be, has held at least since 2012. It is, for all intents and purposes, working.

Stalemate as realm of possibility

If an eventual and total solution to the Diaoyu/Senkaku affair would involve one or more parties losing face, then perhaps, all things considered, it is preferable for the People's Republic of China, the Republic of China and Japan (and possibly third parties like the US) to instead content themselves with rattling sabres for the sakes of their respective citizens.

Moreover, should China or Japan (or both) find themselves in a situation where they feel the need to stoke patriotic and nationalist fervour, then such risky adventures might be best channelled towards a bunch of distant, small and uninhabited rocks. Islands continue to excite and titillate imaginations; and those at the edge of sovereign spaces also serve as putative iconic guardians of the nation (Dodds and Royle, 2003). Concurrently, the evidently circumscribed nature of an island or island group generates a geophysical and mental frame to the whole affair, one that is (hopefully) more easily controlled; and therefore less likely to spill over and infect other spaces. The standstill over the Diaoyu/Senkaku has been quite helpful in this respect to both parties, even if some collateral damage has spilled over from various flashpoints.

Lowering tensions

And yet: could the Diaoyu/Senkaku dispute prove to be that dynamite dangerously close to detonation, even if by accident? Could the situation get out of hand, and serve as the catalyst to serious conflict in the East China Sea, and possibly beyond? The answer to outsiders is an unequivocal 'yes', since such escalations have occurred in recent history, resulting in serious economic and social costs. Indeed, the main loser from the 2012 intensification has been Japan. For the dispute to climb down a few notches of tension and lose dangerous levels of uncertainty, Tokyo and Beijing would need to gravitate towards a behavioural ritual that they can both live with, and seek to lock the matter there. The current situation around the islands approximates this stance. Manoeuvres on the sea (by coastguard vessels warning 'intruders' with the use of megaphones or firing water cannon) and manoeuvres in the air (with scrambling fighter jets intercepting 'incursions' and issuing 'get-out' calls), while preventing nationalist hotheads from swimming to, landing on, and installing anything on the islands in dispute . . . these and similar actions can be construed as pussyfooting. But they can also be perceived as high-stake gambles: given the possibility of human error, something can go wrong:

> Japan's Maritime Self-Defence Force (MSDF) reported that a Chinese frigate locked its firing radar on the Japanese destroyer *Yudachi* on January 30, 2013. Chinese authorities instigated an investigation into the incident in response to Japan's protest, leading to speculation that Beijing was unaware of the ship captain's actions.
>
> (Smith, 2014a, pp. 1–2)

> A fishing captain could take action against a coast guard vessel, activists might attempt to land on an isolated island . . . or a pilot might be ordered to confront an unidentified airplane in overlapping ADIZs, gets a little too close and loses control of his [*sic*] jet fighter.
>
> (Smith, 2014b, p. 129)

And yet, this could be a level of risk that both sides think they can live with. Not only that, but this 'risk-management' approach adopted by the governments in Beijing and Tokyo is being used precisely to legitimate a more muscular style of governance of bilateral relations in both countries (Hook, 2014; Maslow et al., 2015).

The respective leadership of China/PRC and Japan are on a roll. Within just three years, Xi Jinping has skirted away from the collective leadership that has characterised leadership in China post-Chairman Mao. He has consolidated his power at the helm of all control structures in China as 'Chairman of everything', and is now openly referred to as 'the core leader' by senior Communist Party officials (Carney, 2016; *The Economist*, 2016a). Xi's decisions look set to prove that China can no longer 'hide its light' as Deng Xiaoping used to urge, and especially in a 'multipolar' world with no single and undisputed hegemon around (*The Economist*, 2014b).

Across the sea, Shinzo Abe, who is also the President of the Liberal Democratic Party (LDP), served a brief and rather undistinguished term as Prime Minister of Japan (2006–2007) when he resigned for health reasons. But he returned from the political sidelines to victorious re-election in 2012, winning convincingly again in 2014. In 2016, Abe's LDP and its coalition partners secured a two-thirds majority in the legislature's upper house, to go along with their two-thirds majority in the lower house. Such double super-majorities are required to begin the process of amending Japan's constitution, which remains a central plank of the LDP's policy platform (Gelernter, 2016). Abe's foreign and security policy – highly charged with an ideological and historically revisionist discourse – contains the potential to shift Japan onto a more assertive international trajectory, possibly displacing the 'Yoshida Doctrine' that has been Japan's guiding grand strategy since 1945 (Hughes, 2015). Interestingly, the home base of Japanese Prime Minister Abe is none other than Shimonoseki City, in Yamaguchi Prefecture (山口県), southern Honshu, the site of the agreement that closed the First Sino-Japanese war in 1895.

The two leaders thus find themselves in extremely powerful positions in their respective countries, and their commonly hawkish approach to the Diaoyu/Senkaku affair may explain much of their strong and ongoing domestic support.

In this sense, the stand-off over a bunch of uninhabited rocks and islets has delivered considerable political capital to both leaders, and their respective hawkish factions, in China and Japan. Are we then surprised that the impasse continues in much the same way? In politics, a festering island row can be good for you.

14 Speculating

> Once political traction has been re-established, it may be possible to deal with the islands in a manner that befits their actual insignificance. For now, de-escalation is the better part of valour and of national interest. Farther down the road, creative political solutions may become possible. Imagine a future summit in which Chinese and Japanese political leaders . . . *establish the islands as an International Nature and Wildlife Preserve.*
>
> (Przystup and Saunders, 2013, my emphasis)

> [E]xplore the option of shelving sovereignty claims: either by sharing out the resources or *declaring the islands and their waters a marine nature reserve.*
>
> (*The Economist*, 2012a, my emphasis)

Introduction

The status quo may be holding, and all 'solution protocols', proposed and otherwise, may pale in the face of the staying power of the current stand-off in the East China Sea. It is especially so if the circumstances align with the domestic political agendas in both Tokyo and Beijing. Neither of the two leadership cadres of these neighbouring countries may sense that they can earn any brownie points by shifting – let alone improving – cross-sea relations at the present time.

And yet, situations can and do change. One thing is certain: China and Japan are geographically destined to remain neighbours, and important trading partners. One may wish the other to go away; but it will not. Might the respective leadership not grasp at opportunities that could transform a tense situation into one that offers creative and constructive possibilities to both sides? Might Xi Jinping and Shinzo Abe jointly consider a legacy project that could put to rest long-running animosities?

Previous chapters have considered various solution protocols rather factually, without suggesting which one might be better or less suited to handle the Diaoyu/ Senkaku affair. This chapter proposes what can be seen – at least by this author – as an ideal (albeit speculative) win–win scenario, with coastguard vessels currently trading warnings and firing water cannons at each other replaced by cruise ships escorting eco-tourists to explore what could become designated as a trans-border nature reserve or even a UNESCO World Heritage Site, following co-nomination by China and Japan.

Or: drop the cruise ships altogether, and go for a marine protected area co-sanctioned and/or co-managed by China and Japan – and, who knows, even Taiwan? – with the Diaoyu/Senkaku islands ensconced within.

Once again, islands come to prominence: the world's largest single marine protected area (MPA) was constituted by the government of the United Kingdom around the Pitcairn islands in 2015 (Vaughan, 2015). This is one of the latest in a series of MPAs which are increasingly recognised as vital instruments for the promotion of marine conservation and management, for protecting biodiversity, and for supporting the sustainable use of marine species (notably fish but also seabirds) and other marine resources, such as coral and algae (Edgar et al., 2007).

It would not be the first time that a trans-border location, finding itself in the midst of tensions between two suspicious, even hostile, neighbouring countries, ends up in a space and time warp, and from which nature benefits. For years on end, the impossibility of human visitation or settlement in a 'no man's land' allows an unmolested Nature to take the upper hand. A fine example of this is the Korean Demilitarised Zone (DMZ): created by agreement between North Korea, China and the United Nations in 1953, the DMZ is 250 km (160 miles) long, and around 4 km (2.5 miles) wide, and separates North Korea from South Korea (Weisman, 2008). This may be the world's most fortified military border; but 'Wildlife thrives in the "most dangerous place on earth"' (Moss, 2014). Might not the Diaoyu/Senkaku follow suit?

Note: Diaoyu Dao/Uotsuri-shima, the largest island in the group, is known to harbour a number of endemic species. These include the Senkaku mole (*Mogera uchidai*) – with just one specimen identified in 1991 – and the Okinawa-kuro-oo-ari ant (IUCN Red List, 2016; Japanese Ant Image Database, 2016).

History as prologue

The fantasising of such a rapprochement is not just fanciful 'pie in the sky'. Once again, past history is our guide. We have already had a tense stand-off between China and a neighbouring island jurisdiction. And this relationship, while still remaining somewhat tense and fraught with political see-sawing, has matured steadily via various 'confidence-building measures' to support peaceful, 'win–win' actions which include trade and inter-territorial tourism. I am referring to the relationship between China/PRC and Taiwan/ROC. Today's so-called cross-strait relations are a far cry from what prevailed a few decades ago. In a span of almost 70 years, the situation has moved from a bloody civil war with military manoeuvres and incursions, to a gradual thawing of (first unofficial, then increasingly official) relations, hostile non-contact during the previous DPJ government in Taiwan (1998–2008) and the resumption of high-level contacts since (Wang, 2015). The situation remains volatile, especially with the return of a DPJ government in Taipei, but cross-strait relations have much improved. And, in the spirit of this book we need look no further than a small island for the evidence of such a thaw. On Kinmen island, part of the ROC but located within spitting distance of Xiamen city and the PRC mainland, an interesting reversal of function has occurred: recovered artillery shells, vestiges of the bombing campaigns by the PRC on the island, are now crafted into knives to be bought by tourists who, since 2015, come mainly from the PRC; those who wielded guns are now wielding cameras (Chun Han Wong, 2015; Zhang, 2010).

Might not a similar development unfold in the East China Sea?

First of all, Northeast Asia has seen a phenomenal growth in the cruise-ship industry in recent years. Demand amongst Chinese, Korean and Japanese publics for cruise-ship tourism is increasing. Yet the industry still remains largely untapped (Lee, 2016). Tension in territorial waters does not help, but, should regional relations improve, the industry is poised to take off.

Second, one major character in this speculative narrative has been very much a victim of the recent history of the Diaoyu/Senkaku island chain. Could the situation reverse itself, with the hitherto agonist become the protagonist of and for a cruise-ship and/or eco-tourism industry?

Enter the short-tailed albatross

Historically, the short-tailed albatross (*Phoebastria albatrus*) was probably the most abundant albatross in the North Pacific, with potential breeding sites also in the North Atlantic (Olson and Hearty, 2003). However, starting before and after the turn of the twentieth century, millions of these birds were hunted for their feathers, oil and fertiliser (USFWS, 2008). Such indiscriminate hunting occurred also on some of the Diaoyu/Senkaku Islands (see Chapter 4). By 1949, no birds were observed breeding and the species was thought to be extinct. In the US, this albatross species was federally listed as endangered throughout its range in 2000. Today, breeding colonies are known to exist on just two small islands in the western Pacific. Some 80–85 per cent of the breeding population can be found on Torishima (鳥島, literally 'Bird island'), an uninhabited Japanese islet located some 600 km (373 miles) south of Tokyo and which is an active volcano: one major and sudden eruption, therefore, and this main colony could be easily wiped out (USFWS, 2008, p. iii). The other breeding colony is found in the Diaoyu/Senkaku islands (Good et al., 2105, p. 3). While the albatross were massacred there in their thousands in the early twentieth century, the abandonment of the settlements on the islands since the late 1930s and the lingering dispute over the ownership of the islands has meant a long human absence, allowing the albatross to come back, re-settle and rebound.

The return of the albatross signals wider changes on the Diaoyu/Senkaku islands. While the respective claimants engage in their political games and manoeuvres, Nature has not been idle. The islands can be said to have returned to a wild and rambunctious condition, taken over by vegetation, while preserving the relics of an industrial era. The albatross feather and the bonito processing 'factories' have now been lying idle for almost a century, legacies of a bygone age.

The situation is remindful of another set of islands where the human footprint is minimal and nature has been preserved. This set of islands also happens to straddle international borders: indeed, not two countries but three in this case. Amongst these, there is also a legacy of war: two of these countries were occupied by forces from the third during World War Two. Moreover, these islands are inhabited and their resident populations play a part in the management of the land and sea as a special trilateral space. We are referring to the ecosystem of the Wadden Sea, in Northwest Europe: the world's only UNESCO World Heritage Site that consists of islands, and their surrounding tidal flats, shared between more than one country, as well as a major tourism attraction.

Trilateral management of the UNESCO World Heritage Waddensea area and its islands (Jouke van Dijk)

Introduction

The Wadden Sea is one of the world's most valuable stretches of coastline. It extends roughly 500 km along the southeast coast of the North Sea from Den Helder in the Netherlands via the German coast to Blåvands Huk in Denmark. The area contains the world's largest coherent inter-tidal flats: 4700 km^2 emerge at low tide. A large part of the intertidal area is sheltered by barrier islands and sand bars that protect it against the waves of the North Sea. The area includes 25 inhabited islands and several smaller vegetated islands and barren high sands. The total area of the islands is about 2000 km^2 and the Wadden Sea itself covers about 8000 km^2. The Wadden Sea Region is one of the oldest and most complex cultural landscapes in Europe and has been inhabited for more than 5000 years (Knottnerus, 2005; Vollmer et al., 2001). At present, an estimated 3.5 million inhabitants live in the 17,500 km^2 of land that is at risk of severe flooding (i.e. less than 5 metres above and in places even below mean sea-level) (CPSL, 2010), while about 1 million inhabitants live on the islands and administrative regions directly bordering the Wadden Sea. The unique ecosystem of the Wadden Sea is characterised by a rich benthic fauna supporting millions of coastal birds who visit over the course of a year and, due to its dynamic nature, flora and fauna that change over time (De Jonge et al., 1993). The Wadden Sea is also of crucial importance as an internationally shared coastal area for millions of migratory waterbirds in the flyway network connecting the northern parts of the globe with coastal areas across northern and southern hemispheres (Boere en Piersma, 2012).

The unique natural values of the Wadden Sea have been recognised since 2009 when the Dutch and German parts were designated as a World Heritage Site (WHS) and, in 2014, this designation was extended to include the Danish Wadden Sea and, an offshore area, the German Wadden Sea in

Niedersachsen (CWSS, 2008, 2012). It is the only UNESCO WHS that consists of islands and their surrounding waters that are shared between more than one country. However, the WHS is limited to the Wadden Sea so that the Dutch islands and the main navigation channels fall outside. The reasons that the Dutch islands were excluded was due to the fear that the WHS status might limit the expansion of tourism, the major source of income for the islands (Sijtsma et al., 2012). The exclusion of most estuaries forms a real problem because these river-mouth systems may discharge pollution that can directly impact on the WHS. The major German ports of Hamburg, Bremen/Bremerhaven and Wilhelmshaven, the Dutch Eemshaven/Delfzijl and the Danish Esbjerg harbours are all home to important industries.

The functioning of the Wadden Sea ecosystem has been addressed in numerous studies (e.g. Reise et al., 2010; Reise, 2013; Wolff, 2013). There is a rich literature on the governance and management of the Wadden Sea (Kabat et al., 2012) but many of these studies focus on the management of its ecological system only. Although Integrated Coastal Zone Management (ICZM) acknowledges that, in a coastal zone, not only the ecosystem but also the resources and human activities are important, the motivation for the latter's inclusion is that these human activities are seen as a main cause of the disruption of the natural coastal systems. Much less attention is paid to integrated forms of coastal zone management that target balanced economic development to provide future opportunities for people living and working there, while also generating surpluses for environmental protection measures (see, for example, de Jonge et al., 2012). A specific problem for the WHS Wadden is the location in three different countries with each having its own specific national governance structure, in combination with the higher-level regulations from the European Union to which all three countries belong.

To shed light on the complex trilateral governance and management issues of the WHS Wadden, the following section will explore the historical background of the area and the changing perspectives on governance and management related to this.

Historical background

On the large islands of the Wadden Sea Region, there appears to have been almost uninterrupted habitation from the Neolithic. From the sixteenth century, the regional economies increasingly became part of a world economy. With its flourishing agricultural and maritime cultures, this can be seen as the period of greatest economic prosperity for the Wadden region. Significant innovations were introduced to sea fishing in this period, with whaling beginning to flourish from the beginning of the seventeenth century.

In the early modern period, the hunting of harbour seals, grey seals, dolphins and harbour porpoises was a further important source of income, particularly for island and coastal inhabitants. At the start of the sixteenth century, the Frisian coastal areas lost their political independence and were incorporated into larger, centrally controlled territorial states. In practice, because of the remoteness from centres of power, authority was usually delegated to officials, most of whom came from old noble families (Bazelmans et al., 2012).

The historical political development still carries implications for the management of the Wadden Sea Region, especially in the border areas between the Netherlands and Germany and between Germany and Denmark. An interesting issue is, for example, that Germany and the Netherlands for many centuries disagreed about where exactly their shared nautical border lies in Dollart Bay. The territorial dispute has not received much attention for a long time, but has gained relevance as there are plans for major investments in an offshore wind farm, and there are also important economic implications for fishing and the extraction of natural gas in the area. Until today, the problem has not been solved, but to avoid practical problems, in 2014 ministers signed an agreement that the border will remain ambiguous and both nations will share responsibility for the area.

The Dutch island of Schiermonnikoog is the most eastern island from the border of Germany. Since the Middle Ages, it formed part of the Dutch province of Friesland. Its name derives from the grey habits of the Cistercian monks of nearby Klaarkamp monastery on the mainland. When Friesland went over to the Reformation, the island was confiscated by the provincial authorities (1580) and sold in 1638 out of budgetary problems to the noble family Stachouwer. Since then, it has been private property – though administratively remaining part of Friesland – up till the end of World War Two, when it was confiscated by the Dutch government from its last private owner, the German family Von Bernstorff-Wehningen, who had owned the island since 1892.

The most western and largest German island of Borkum has always been a part of the German county of East-Frisia (Ostfriesland); the county itself becoming part of Prussia since 1744. After a short period, East-Frisia formed part of the Napoleonic Kingdom of Holland (1807–1810), the French empire (1810–1813), again Prussia (1813–1815) and the Kingdom of Hanover (1815–1866) when a war between that kingdom and Prussia led to the integration of Hanover (and East-Frisia) into Prussia which in itself became the most influential part of the German Empire. East-Frisia since 1866 has ceased to exist as a separate administrative unit. It nowadays forms part of Regierungsbezirk Weser-Ems, itself divided into so called Landkreise: Aurich, Leer, Wittmund and the city of Emden. Borkum, meanwhile, forms part of Landkreis Leer. Because the geographical location of Borkum is above the Dutch province of Groningen, in 1976 a German company started a successful ferry run from the Dutch Eemshaven, because of the much shorter time needed than from the German mainland.

Also between Denmark and Germany the border has moved back and forth over time. Schleswig and Holstein are the so-called Duchies which, since the fourteenth century, formed part of the Kingdom of Denmark. As such they have a rather complicated history. Schleswig is located north of the river Eider, the traditional northern frontier of the medieval German empire. It has a largely German- or Frisian-speaking population, already since 1085 when it was an appendage of the Danish kingdom. Schleswig and Holstein more or less form a unity, since in 1386 the count of Holstein obtained Schleswig as a fief from the king of Denmark. The situation became even more complicated when, shortly after the death of King Christian I in 1481, his two sons and their heirs divided the Duchy of Schleswig and eventually Holstein too into a royal (Segeberger) and a ducal (Gottorper) segment (1490, 1544/1581). The northernmost Wadden Islands of Fanø and Mandø have always formed an integral part of the Kingdom of Denmark. This also goes for the southern half of Rømø and (as far as the North Frisian Islands are concerned) for the northern tip of Sylt (List and surroundings), the whole of Amrum as well as the western half of the isle of Föhr. The other islands were under the purview of Schleswig-Holstein-Gottorp, but enjoyed a lot of autonomy.

In 1864, Prussia declared war on Denmark and annexed the Danish Duchies, including the islands of Romø and Fanø, up till the Kongeå river; the new frontier leaving the town of Ribe just a handful of miles within the Danish border, depriving it from its hinterland. Before that, Romø and Fanø and also Sylt were Danish (Schleswig-Holstein). The loss of their North Sea ports like Ribe, Husum and Altona prompted the Danish government into the development of a new port on its western coast: Esbjerg from 1868 onwards. After World War One, a plebiscite among the population of Schleswig divided this former duchy into Danish and German parts. Denmark reclaimed Rømø and Fanø in 1920.

Overview of the coastal management strategies and goals in the Wadden area

There is a rich literature on the governance and management of the Wadden Sea (Kabat et al., 2012; Van Dijk et al., 2016). An important aspect of the governance and management is defining the goals to achieve. Over the centuries we can see major changes. Safety and protection against flooding have historically always been major issues in the Netherlands. The first large-scale dyke-building and land-reclamation work started as far back as the eleventh century, and this changed the hydraulics and related processes such as water transport and erosion and sedimentation patterns in the Wadden Sea (Bazelmans et al., 2012; de Jonge, 2009; Reise, 2005). When people behind the dykes began to feel safe, it also marked a significant change in the relationship between man and nature. This may have been a prerequisite for the perceptual shift from a threatening sea to a Wadden Sea that has become increasingly threatened by human impacts and consequently also deserving of protection from humankind (Fischer, 2011). The largest project to date has been the separation of the Zuiderzee (now the IJsselmeer) from the Dutch part of the Wadden Sea in 1932 by the construction of the Afsluitdijk to improve coastal protection and to gain agricultural land in the Netherlands. This completely changed the ecosystem of the former estuarine system and reduced the populations of diadromous fish (those that can live in both fresh and sea water) (De Beaufort, 1954; de Jonge et al., 1993). The almost one thousand-year tradition of reclaiming land from the sea for agriculture slowly ceased in the second half of the twentieth century as nature became more highly valued than economic gain (Wolff, 1992). The solid and visionary work of the Mazure committee led to the abandonment of plans for the total enclosure of the former Zuiderzee and the Wadden Sea, with dams between the barrier islands and across estuaries (Waddenzeecommissie, 1974). A cost–benefit analysis showed that reclaiming the land was not even economically profitable (Oosterhaven, 1981). Similarly, land reclamation for coastal protection in Germany came to an end in the 1980s.

As land reclamation stopped, there were growing concerns about the impacts of other human activities such as fisheries, hunting, military exercises, coastal protection, tourism and industrial development. These concerns led to the implementation of several policy measures to protect the ecological values of the region. In the new millennium, the discourse shifted from protecting nature towards one where nature protection and enhancement could go hand-in-hand with sustainable economic development (de Jonge et al. 2012; Reise, 2011; Runhaar et al., 2010). The turning point was the report by the Advisory Group on Wadden Sea Policy (Meijer Committee) published in 2004 (Adviesgroep Waddenzeebeleid, 2004). The committee found that the policy and management style adopted had been defensive, and that this had unintentionally resulted in the impeding of the development and improvement of the natural environment, and that the economic development of the Wadden Sea Region and the northern part of the Netherlands had been blocked. The committee believed that an integrated perspective based on prioritising the natural environment but with limited and shared human use was necessary in order to monitor and continue to develop the values and interests at stake, preferably at a trilateral level.

Although the WHS Wadden comprises three countries, the government and management is still very complex and far from integrated due to the difference in the national governance structures and the overarching EU regulations. The German and Danish authorities have chosen a relative simple management model via the status of National Parks. The advantage of this approach is that rules are made at the central level and there is only one governing body. In contrast to this is the governance of the Dutch Wadden region which has always been complex and fragmented (Rekenkamer, 2014). In the Netherlands, five national ministries are directly involved in the governance of the Wadden region, together with three provinces, four water boards and 18 municipalities (Toonen and Staatsen, 2004). In addition to the governmental authorities, other groups play a role, such as private property managers, NGOs and a variety of private organisations. Good governance requires an important role for non-governmental parties, civil society and market players. A number of discussion groups and coordinating groups exist in order to care for the extensive horizontal and vertical coordination between the different authorities and stakeholder groups. Research on governance of the Wadden was conducted by Klostermann and Toonen (2011), Toonen and Staatsen (2004) and Klostermann et al. (2009), among others.

Since 1996, the European Commission has intensively worked to promote Integrated Coastal Zone Management (ICZM) as an approach to integrated planning and management, in which all policies, sectors and interests are properly taken into account to achieve sustainable coastal development. The EU ICZM Recommendations, adopted by the Council and Parliament in 2002, requested the EU Coastal Member States to set up strategies to promote ICZM along their shorelines. ICZM can be regarded as an approach to good territorial management including land–sea interactions. As such, the European developments on ICZM are of relevance to the governance of the Wadden region. The most important pieces of EU legislation at the moment are the Birds and Habitats Directives, Environmental Impact Assessments, and the Water Framework Directive (Steijn and Huizer, 2013)

On top of national and EU regulations, it was being increasingly felt already for many decades that, in addition to regional or national policy measures, a trilateral approach was also needed in order to protect the ecological values of the Wadden area. As far back as 1974, the Dutch government was consulting with its counterparts in the Federal Republic of Germany and Denmark about international cooperation to protect the Wadden Sea. These consultations ground to a halt due to disagreements between the federal and the regional state governments in Germany. To resolve the problem, it was agreed that scientists of the three countries should investigate the matter and this resulted in a report (Wolff, 1976) on the desirability of international Wadden Sea protection. In 1982, a Joint Declaration on the Protection of the Wadden Sea was agreed in which the three countries declared their intention to coordinate their activities and measures for the protection of the Wadden Sea. To facilitate this cooperation in nature management, a Trilateral Wadden Sea Cooperation (TWSC) was established in 1987, with a Council of Ministers that meets every three years at the Wadden Sea Governmental Conferences plus a Wadden Sea Board (WSB) that is the governing body of the Cooperation that supervises the Common Wadden Sea Secretariat (CWSS). In 1997, a Trilateral Wadden Sea Plan was adopted that defined shared management targets and this was updated in 2010 (CWSS, 1998, 2010).

Following a decision by the 9th Governmental Conference of the Trilateral Wadden Sea Cooperation in 2002, the Wadden Sea Forum (WSF) was established. Although the WSF (2010) has a broader scope than the TWSC, the two organisations are working together in many fields and they share a building in Wilhelmshaven. Nevertheless, the WSF is constituted as an independent platform of stakeholders from Denmark, Germany and the Netherlands to contribute to an advanced and sustainable development of the trilateral Wadden Sea Region. In particular, this involves integrating specific environmentally sound, economically viable and socially acceptable cross-sector and trans-boundary strategies, actions and techniques. The WSF includes representatives from agriculture, energy, fisheries, manufacturing, harbours, nature protection and the tourism sectors, as well as from local and regional governments. National governments are present as observers (WSF, 2010).

Whereas the TWSC has focused mainly on the conservation of ecological values within the Wadden Sea, the WSF has a broader goal: to help achieve a sustainable society by 2030 in which economic activity supports social development and safeguards healthy ecosystems and culturally historic landscapes throughout the Wadden Sea Region. The Wadden Sea Forum has adopted integrated coastal zone management (ICZM) as a process for achieving sustainable development of the Wadden Sea Region. The increasing uses made of the offshore Exclusive Economic Zones (EEZs) beyond the protected Wadden Sea have led the WSF to enlarge the spatial scale of its work regarding ICZM objectives. Thus, the World Heritage Site (WHS) property is restricted to the Wadden Sea (CWSS, 2016), whereas the Wadden Sea Region covers a much larger area and is more relevant to an integrated management approach, embracing both ecological and economic perspectives (Wadden Sea Forum, 2016). In particular, this management involves integrating specific cross-sectoral and trans-boundary strategies, actions and techniques in order to facilitate the coordinated growth and implementation of trilateral strategies and action plans with regard to spatial planning, infrastructure, transport, agriculture, fisheries, tourism, coastal defence, shipping, port facilities, social services, nature protection and any other policy areas that have a possible impact on the sustainable development of the Region. The related ICZM strategy is documented in the ICZM Strategy for the Wadden Sea Region (WSF, 2013).

The WSF ICZM strategy aims to achieve a balanced population structure with a healthy share of young and elderly people, with indigenous young people choosing to stay in the region and further population decline being avoided. The WSF also strives for near full employment which requires a wide diversity of jobs for residents ranging from high-level professionals to those seeking manual employment. In terms of welfare, WSR residents should have adequate income, good living conditions and a high life expectancy. Social equality and a good work–life balance should ensure a satisfied population. All this requires a balanced economic development that provides future prospects for people living and working in the area, but one that should also generate surpluses for environmental protection measures. The economy in the WSR should be designed to be resilient to economic recessions and international competition through the adoption of strategies for innovation and entrepreneurship.

This is outlined in more detail for specific sectors (WSF, 2013). As an example of progress, the ministers in the Wadden Sea Council meeting at the 2014 Trilateral Governmental Conference in Tønder adopted a joint strategy for 'Sustainable Tourism in the Wadden Sea World Heritage Destination' (CWSS, 2014). This strategy has been developed at the request of the World Heritage Committee using a participatory approach, and a solid transnational network of stakeholders has been established over the last two years that will provide the framework for implementing this joint strategy. In terms of key economic activities, tourist-related activities are linked to the unique assets of the Wadden area. On the islands, most of the tourist activities are located on or near to the North Sea beaches rather than on the Wadden Sea. Nevertheless, the islands carry the burden of an increasing number of tourists with impacts on land use, pollution, the local population, and so on. On the German island of Sylt, for instance, the inflow of rich and famous people caused an enormous increase in housing prices with the result that most local islanders are forced to live on the mainland and commute back to the island for work every day. Tourists and all the associated supplies need to cross the Wadden Sea to the islands, which again requires navigable channels that involves regular dredging and thus potentially harmful activities to the Wadden Sea. On the mainland coast, there are substantial tourism activities in the German section of the Wadden area, but not in the Dutch or Danish parts.

Remarks

ICZM, as defined by the WSF (2013), aims to manage the Wadden area such that both the ecosystem and the social system can develop sustainably. Humans occupy and use the area, and consequently put direct pressure on the ecosystem (such as through fisheries) and also apply indirect pressure to its physical (such as through dredging and channel deepening) and chemical subsystems (such as through pollution) (de Jonge et al., 2012). The effects of this relationship are many and complex and, consequently, the system needs to be managed so that both the ecosystem and the social system can develop in a sustainable way. This in itself is a real challenge for integrated coastal zone management in the broader sense. Proper management can only be applied if there is sufficient ecological and socioeconomic information and knowledge to manage this integral system (De Jonge et al., 2012; Giebels et al., 2013; Kabat et al., 2012). In addition to scientific knowledge, participatory knowledge and practices are also increasingly required to facilitate the integration of different types of knowledge, both formal and informal, and the inclusion of a range of stakeholders to enhance the ICZM (Puente-Rodríguez et al., 2015). It is clear that knowledge is a key feature in achieving ICZM, but also that the decision-making process is important.

While there are many ideas stemming from environmental disciplines on managing the Wadden Sea as a natural system, there is a real knowledge gap when it comes to its economics. A recent overview addressing many aspects, but not the economy, can be found in a volume published by *Ocean & Coastal Management* (Van Leeuwe and Haartsen, 2012). The studies that do offer insight into the socioeconomic situation and developments in the trilateral Wadden Sea Region are sparse and not very recent. Arndt et al. (2004) and Van Dijk et al. (2016) are exceptions. These authors present figures on demographic developments, the regional economic structure and the labour market over the period 1990–2000. These studies indicate that the massive demographic shift due to population decline and aging will have a strong social and economic impact on the region. The regional economic structure will change from manufacturing to services, and also within manufacturing and services major changes will take place. Employment is expected to continue to fall in agriculture and fisheries. Nevertheless, along with tourism, these sectors do provide employment where often other job opportunities are limited, resulting in high unemployment in several sub-regions. These authors also stress that the Wadden Sea Region is heterogeneous in being situated in three countries and various sub-regions with different local foci and that, therefore, future strategies will have to address the regional differences between the sub-regions. Social equality and a satisfactory work–life balance should be achievable by Wadden residents as much as by those living elsewhere. This requires ICZM to aim for balanced economic development that provides prospects for people living and working in the Wadden area, but that also generates surpluses for environmental protection measures to preserve the unique natural values of the Wadden area since these are a major asset in creating jobs in tourism, and also provide clean air and a healthy living environment.

Based on these pressures and the management challenges inherent to a broader ICZM approach, insights are needed into the combined ecological and socioeconomic system of the whole trilateral Wadden Sea Region in order to design science-based monitoring and management support systems. The essential characteristics of this approach are strongly interdisciplinary and require an ICZM focus on aspects of scale and cumulative processes to keep track of the impacts of autonomous processes and human activities on the unique natural values of the Wadden UNESCO World Heritage Site. This management becomes even more difficult due to the complicated governance structure resulting from the fact that the Wadden Sea Region includes 25 inhabited islands and their surrounding waters are shared between three countries. Although the countries have very friendly relations and have already been developing joint policy initiatives for several decades, the management of the WHS is still fragmented and not very effective due to the complicated differences in governance structure between the countries; on top of that, EU regulations also play a role.

The Wadden Sea is an ecosystem that happens to straddle a trilateral border. The three countries involved – Denmark, Germany and the Netherlands – have deemed fit to develop intricate systems of co-management over the site, as the above text clearly illustrates. The UNESCO World Heritage Site seal of approval is both the cause for, and effect of, such intensive (and peaceful) cooperation. Similar initiatives could yet unfold in the East China Sea. And lest someone remind us that there is so much historical memory, and therefore no love lost, between China and Japan as to make any *rapprochement* virtually unthinkable, then let us remember also that Germany occupied both Denmark and the Netherlands during WWII; and that both Denmark and the Netherlands have historically experienced tense moments with their much larger and stronger neighbour. Of course, their common membership in the European Union (EU) – Germany and the Netherlands as founder members of the European Economic Community (EEC) in 1958; Denmark joining in 1973 – helps. The closest equivalent to the EU in East Asia is the Association of South-East Asian Nations (ASEAN); but this is not as strong a politico-economic bloc as the EU; moreover, ASEAN only holds 'dialogues' with both China and Japan, since neither of these two countries are members.

Examples of hope

So: where does this leave us?

In this text, no less than seven so-called 'solution protocols', all involving islands, have been proposed and reviewed, each of which, singly or in combination, *can* offer opportunities for contemplating 'win–win' outcomes for the Diaoyu/Senkaku islands in the East China Sea. Some of the island cases deliberated upon may have, *prima facie*, presented themselves as even more intractable than the

Diaoyu/Senkaku kerfuffle. At the same time, we have soberly commented that the 'status quo' offers benefits to the two parties. Moreover, the situation is not helped by Tokyo officially refusing to declare that there is a dispute over the islands; as well as officially refusing that there was any agreement with China to 'postpone' or 'shelve' the putative issue.

And so, to conclude, a review that has so far been missing in this book is to take a closer look at, on the basis of past practice, how well disposed China and Japan may be to consider any such solution protocols. One may hazard to say that, if the two countries have already had some experience of and in 'creative governance', then there may be some grounds for guarded optimism even in the East China Sea.

China

Here, China/PRC comes out as the better experienced of the two parties. Its larger size and diverse population, its experience with European colonial powers and its long land border with so many neighbours have all led to a range of accommodations that challenge orthodoxy. Moreover, and unlike Japan, some of these 'accommodations' are not lost in history but are ongoing.

China is a centralised yet federative entity: it is organised in terms of various administrative divisions which, however, do not enjoy the same status. While the 'normal' unit of subnational division is the province – of which there are 22 – there are also five 'autonomous regions' on the country's landward periphery, mainly respectful of ethnic and national minorities. There are also four large metropolitan areas, giving four mega-city municipalities (Beijing, Shanghai, Tianjin and Chongqing) special governance powers.

As consequences to negotiations leading to the end of the European presence on the Chinese coast, both Hong Kong (handed back by Britain in 1997) and Macau (handed back by Portugal in 1999) are special administrative regions (SARs) with the most considerable autonomy within the contemporary Chinese state; although this *détente* is understandably not always easy, and often messy. We should add the island jurisdiction of Taiwan which, from the perspective of Beijing, is the renegade '23rd' province of China, but which has been offered an SAR status should it willingly rejoin, or be forcibly rejoined to, the (patient) motherland.

Before that, various parts of China had the peculiar privilege of having constituted international settlements on Chinese soil. For almost a century (1757–1842), the port of Canton – now Guangzhou – was the sole legal site for all Chinese trade with the West. Here, Shamian Island was the key centre of operations: foreigners lived, stored goods and did business from a row of houses known as 'The Thirteen Factories'. The island was connected to the mainland by means of two bridges, which were closed at night for security purposes (Dorow, 2006). Then, the Shanghai International Settlement emerged after the defeat of the Qing dynasty by the British Empire in the First Opium War and the subsequent Treaty of Nanking in 1842 (Chapter 3). Under the terms of this treaty, the Chinese city of Shanghai, along with four treaty ports – Canton, Ningbo, Fuzhou and Amoy (now Xiamen) – were obliged to open up to foreign trade. Just over 80 treaty ports were established in China, accommodating many foreign powers. These included the city of Weihai, part of territory leased to Britain, and known as Port Edward during this time.

Finally, swapping territory can be expected were China to show resolve to conclude even some of its land disputes, particularly the ones with India. Where river islands have been involved in land disputes, we have already commented on the readiness, *in extremis*, of Beijing and Moscow to break with convention and split the single island of Heixiazi (in Chinese)/Bolshoi Ussurriskii (in Russian) on the Amur River (Chapter 7).

All this experience may yet come into play as China, one may portend, eventually settles its remaining land-border disputes with Bhutan, India, North Korea and, who knows, perhaps even the Paracels, the Spratlys and other contested islets, atolls and reefs in the South China Sea: 'Despite loudly spoken, nationalist assertions not to cede "an inch of Chinese territory," the PRC's practices on its many, lengthy borders show a great deal of flexibility and innovation' (Lee, 2015, p. 1).

Japan

In contrast, Japan has had much less of an obligation to craft or acknowledge 'diversity in unity' within its borders. Japan is essentially a bi-cameral but unitary state, containing 47 administrative divisions (or prefectures), and with the Emperor as its titular head of state. Local jurisdictions, or municipalities, depend considerably on the national government both administratively and financially. The 47 prefectural units include one metropolis (Tokyo), two urban prefectures (Osaka and Kyoto) and one territory (the island of Hokkaido). The Ainu, mainly in Hokkaido in the extreme north, were only recognised as 'an indigenous people with a distinct language, religion and culture' in 2008 following a parliamentary resolution. Japan has not had the pressures and exigencies imposed on China to consider creative governance options: it has no land borders; it has no created differential administrative powers; and the official narrative – the Ainu excepted, but only recently – speaks of homogeneity in terms of race. Indeed, Japan maintains a strict, anti-immigration policy and accepts few refugees, in spite of international burden-sharing responsibilities (Dean and Nagashima, 2007).

Historically, there are at least three examples of 'creative governance' that have featured Japan in its dealings with foreign or international powers. Since we are dealing with modern Japan, they are all island cases. None of these examples persist into the present time.

The first, of course, is *Okinawa* and its islands, under US administration from the end of WWII in 1945 until its reversion back to Japan in 1972 (see Chapter 3). Judging from the reversion agreement, this arrangement included the Diaoyu/Senkaku archipelago.

The second is the island of *Sakhalin*, known to the Japanese as Karafuto (樺太). From 1855, the island was under joint Russian and Japanese control. Then the 1875 Treaty of St Petersburg ceded all of Sakhalin to Russia. But, thanks to Japan's victory over Russia in 1905, the island was split, divided at the 50th parallel by the

Treaty of Portsmouth. The northern section remained Russian, while the Japanese colony of Karafuto was re-established on its southern flank. Russian soldiers overran the southern part of the island in the final days of WWII: after the end of hostilities, its Japanese and Ainu people were forcefully repatriated to Japan. Tokyo has officially given up all claims to Sakhalin (Paichadze and Seaton, 2015).

The third is *Dejima* (出島, 'exit island'). Today, it is difficult to imagine the huge cultural impact of the arrival of the Dutch and Portuguese to an isolationist Japan in the sixteenth and seventeenth centuries. The potential for social, economic and political unsettling that could be caused by the massively different and powerful Europeans obliged the Tokugawa Shogunate of the time to restrict their access savagely: Westerners could only sail into the port of Nagasaki, and to reside on a small and custom-built artificial island, linked to the rest of Japan by means of a narrow bridge, easily guarded and supervised. For over 200 years (1641–1856), the tiny island of Dejima was the only official window of interaction between Japan and the Western world. Its residents were not permitted to wander across the bridge and mingle with the rest of Japan; only a handful of Japanese could secure permission to cross the bridge onto Dejima and 'trade' with the *gaijin* (foreigners) (Kodansha, 1997). Just here, and for two centuries, was an institutionalised 'enclave of difference' (Willis, 2009).

Conclusion: some optimism?

Nevertheless, some optimism is warranted, since some agreements have been struck by China and Japan and also involving their common neighbour Korea. Three joint fisheries agreements have emerged following the ratification of the UNCLOS-related declarations of EEZs by China, Japan and the Republic of Korea and the resulting overlapping maritime claims. The China–Japan agreement of November 1997 relates to part of the East China Sea; the Japan–Korea agreement of January 2000 is in respect of parts of both the East China Sea and the Sea of Japan (known as the East Sea to Korea); while the China–Korea agreement of June 2001 deals with parts of the Yellow Sea. Granted, these joint agreements cover only part of the substantial areas of the overlapping maritime claims of China, Japan and Korea. They are of a provisional nature and without prejudice to any final maritime boundary delimitation. They also have drawbacks, including: they provide for enforcement on a flag state basis while envisaging minimal joint enforcement; and they lack provisions for enforcement against third parties (such as Taiwan which is a significant fishing entity in the waters concerned). Furthermore, they encompass only part of the picture and sidestep substantial 'current fishing patterns': zones where fishing is, at the least, uncoordinated and, at the worst, largely unregulated. Nonetheless, they do represent a positive step towards cooperative, joint solutions to shared problems; they embody a potentially useful application of maritime joint development (Schofield, 2014).

15 Conclusion

Opportunities

In a statement made at the Japan–China Summit Meeting (Prime Minister Takeo Fukuda and Vice Premier Deng Xiaoping) during negotiations on the Treaty of Peace & Friendship between Japan and People's Republic of China on 25 October 1978, Vice Premier Deng Xiaoping made this statement:

> There exist a variety of issues between our two countries. For example, there is the issue of what is called the Diaoyu in China, and the Senkaku Islands in Japan. There is no need to raise subjects like this at a meeting like this. As I expressed to Foreign Minister Sonoda in Beijing, there is probably insufficient wisdom to resolve the issue in our generation, but with the next generation likely to be savvier than us, they will probably be able to find some resolution to the issue. It is essential to look at this issue with a broad perspective.
>
> (Ministry of Foreign Affairs of the PRC, 2014)

Is the 'wisdom' hoped for by Vice-Premier Deng now available? Or is twenty-first-century diplomacy as, or more, vulnerable to patriotic pursuits than that of the twentieth century?

It has been suggested that the cool-headed decision of China and Japan in the 1970s to shelve contentious issues – such as Japan's wartime record and the question of sovereignty over the Senkaku/Diaoyu Islands – appears to have been overturned of late (Choong, 2014). A clear escalation has been noted since the Japanese government went ahead with the purchase of three (hitherto leased) islands within the Diaoyu/Senkaku group in September 2012, involving protests, violent actions and a boycott of economic products. The PRC announced the introduction of new air-traffic restrictions in November 2013, by creating the East China Sea Air Defence Identification Zone (ADIZ) covering most of the East China Sea. The move has been interpreted as an attempt by the PRC to pressure Japan into acknowledging that there is a dispute between Tokyo and Beijing over ownership of the Diaoyu/Senkaku chain (BBC News, 2013). Meanwhile, the Abe government in Japan, via its Foreign Ministry, has of late been displaying a map of the country that proclaims its 'inherent territory' – 固有の領土 (Koyū no ryōdo) – going blatantly beyond its currently internationally recognised borders: the map incorporates three sets of islands – Takeshima, the Northern Territories and the Senkaku – even though these are all claimed or occupied by neighbouring countries: Korea, Russia and China respectively (Dudden, 2015; Ministry of Foreign Affairs, Japan, 2014).

But there are other initiatives that have helped to dampen this dangerous, 'tit for tat' escalation. Already in March 1969, Japan began protracted negotiations with Taiwan and South Korea, leading to an agreement in principle in September 1970, to set up a joint development project in the East China Sea (Drifte, 2009). The South Korea–Japan Joint Development Agreement was signed in 1974. Moreover, in 2013, after 17 years of negotiation, Japan and Taiwan signed a 'civil' fisheries agreement. The agreement covers the waters around the contested islands which currently fall within overlapping jurisdictional claims (Taiwan–Japan Fisheries Agreement, 2013); this is done fully within the spirit of Paragraph 3 of Article 74 of UNCLOS (1982). Under this landmark deal, Japanese and Taiwanese fishing boats in the Diaoyu/Senkaku island area are exempted from the jurisdiction of each other's law enforcement, and the two sides agreed further to discuss resource conservation and common fishing rules. This deal was deemed a 'strategic success' for both Taiwan/ROC and Japan, to the consternation of China/PRC (Kotani, 2015).

Can such agreements, and their spirit of pragmatism, extend to other areas of joint exploration or management between Japan and China, and possibly dissipating the stubborn irritant that has become the Diaoyu/ Senkaku island affair? Surely, there is no shortage of examples from which to either choose or 'mix and match'.

The ROC/Taiwan government, on its part, has proposed the East China Sea Peace Initiative which, while reiterating (but side-stepping) claims to sovereignty over the Diaoyutai/Senkakus, augurs towards 'shelving disputes, pursuing peace and reciprocity, and promoting joint exploration and development' (Ministry of Foreign Affairs, ROC, 2012).

When meeting Suzuki Zenko, an LDP member of the lower house of the Japanese Diet on 11 May 1979, PRC Vice-President Deng Xiaoping said that consideration may be given to joint development of the resources adjacent to the Diaoyu islands without touching upon the touchy question of its territorial sovereignty. In June 1979, the Chinese side formally proposed the concept of joint development of resources adjacent to the Diaoyu islands to the Japanese side through diplomatic channels (Ministry of Foreign Affairs of the PRC, 2014).

Of late, after a serious escalation phase, China and Japan have sought to reduce the tension over the islands, particularly when confronted by urgent priorities to restore and boost their mutual economic links and trade (Perlez, 2014).

Reflections

Some solution protocols have worked in some cases; others have not. Some tested protocols can be repeated; others may not qualify. We recognise the problem of looking at *existing* disputes: they act as self-righteous paradigms, and often seem to defy resolution. Then there are also the pros and cons of *non*-dispute resolution: in this case, shelving the issue, even in the name of regional peace and cooperation, officially or otherwise. This stance, however, can be seen to work mainly in favour of China, which continues to grow stronger militarily and economically (and against Japan, which may appear to be a regional behemoth on the wane).

Islands have been 'novelty sites' in more ways than one (Baldacchino, 2007): they have served as laboratories for evolution, nurturing high levels of endemism (Whittaker and Fernández-Palacios, 2007); they have encouraged contemporary offshore pursuits (Palan, 2006) and so many other examples of playful, 'creative governance' (Baldacchino, 2010). Cannot this track record be better acknowledged, and extended to dispute resolution? Can the parties with interests in this issue look beyond anticipated zero-sum solutions as can be the outcome of international litigation, as in the 1928 Island of Palmas award (Stubbs, 2014)? An international nature and wildlife preserve, or a marine protected area, for the Diaoyu/Senkaku islands have been suggested as a 'dream solution' (Przystup and Saunders, 2013) and can serve as that elusive middle ground; this book echoes these possibilities. And, just as Okinawa's future has been suggested (by Professor Sato Manabu) as an 'open border land' (McCormack and Norimatsu, 2012, p. 219), the same may be suggested for the Diaoyu/Senkaku chain.

Closing remarks

This book's approach is not new. It has followed a path taken by others in grappling with other challenges: consider Elinor Ostrom and her eight 'design principles' for common property regimes which militate against the 'tragedy of the commons' (e.g. McKean and Ostrom, 1995); or Jim Dator and his four 'archetype solutions' to nagging disputes (Dator, 2009). The road ahead would also involve confidence- and security-building measures (CSBM) to at least start considering these alternative options: 'Even those who recognize the importance of transcending such logic in favour of condominium or shared sovereignty options, generally hold that the issue can only be addressed at the last stage of any negotiated process, if at all' (Drifte, 2014, n.p.). The security of, and any conflict in, East Asia has tremendous implications for global security: no wonder then that the Diaoyu/Senkaku affair has been the subject of considerable texts, expert reviews and opinion pieces.

What is new and original to this book, however, is the explicit rendering of islands as *focus* – and not just islands as *locus*, the coincidental material and geographical platform of what is unfolding (Ronström, 2013). To echo a UN phrase from the 1st Small Island Developing States (SIDS) summit in Barbados in1994, these are not just island matters: *islands matter* too (Hills, 1996). A closer look at the circumstances and processes that brought contending regional powers to an honourable, even mutually advantageous, settlement over *other* islands would be one timely and valid approach with respect to the festering dispute in the East China Sea.

The sheer minor scale and size of small (and especially uninhabited) islands and archipelagos, and their often lack of exploitable, land-based natural resources, may create those very conditions conducive to playful experimentation between contending parties, leading to creative solutions, and which challenge conventional conceptualisations of sovereignty.

Epilogue I

The Diaoyu/Senkaku islands dispute and future development of Sino-Japanese relations

Liu Jiangyong

Introduction

The year 2012 witnessed the 40th anniversary of the regularisation of Sino-Japanese diplomatic relations, the 640th anniversary of the conferment of the title of King of Ryūkyū by Hongwu, the first emperor of China's Ming dynasty; the 420th anniversary of Japan's invasion of the Korean peninsula under the leadership of Hideyoshi; the 140th anniversary of the reconfiguration of the Ryūkyū Kingdom as a feudal domain (*han*); and the 120th anniversary of Japan's completion of preparations for the Sino-Japanese War of 1894–1895. The major historical events mentioned above can all be seen as directly or otherwise related to the lingering Diaoyu islands – Senkaku islands in Japanese – dispute between China and Japan.

The Diaoyu island dispute was officially shelved with the normalisation of Sino-Japanese diplomatic relations in 1972, but it has always been there as a nagging source of anxiety and concern. When the Noda government in Japan announced that it was 'purchasing' three of the Diaoyu islands in 2012, this decision highlighted the unsettled nature of the matter and impacted heavily on the overall tenor and state of Sino-Japanese relations, with potentially dangerous consequences.

To correctly understand and properly handle the current Sino-Japanese dispute over the Diaoyu islands, one first needs to learn from historical experiences and lessons, so as to make more accurate pre-judgements and policy choices for the future.

The decision to 'purchase' the islands

On 10 September 2012, the Noda cabinet in Tokyo announced that it had 'purchased the islands', marking a low ebb in Sino-Japanese relations. This decision was taken in response to the intention expressed by Shintaro Ishihara, then Governor of Tokyo, to purchase these islands: mindful of the serious consequences of allowing such a transaction, the Noda government pulled the carpet from under Ishihara's feet by effectively 'nationalising' the islands. In essence, this move was an attempt to achieve the long-term stable and secure administration of the islands. The Noda government decision was taken even though PRC President Hu Jintao had told

Noda in person that Sino-Japanese relations were under serious threat because of the Diaoyu affair, and that Beijing was resolutely opposed to any attempt by Japan to purchase the islands, a move that would be considered illegal, invalid and provocative by Beijing.

Meanwhile, in July of that same year, then Prime Minister Noda had also suggested that Japan would consider dispatching its self-defence forces if any other country invaded 'Japanese territory and territorial waters', including the Diaoyu islands and surrounding waters. Had such decisions unfolded and such a trend continued, the serious of events would have brought serious and unimaginable consequences for Japan and Sino-Japanese relations generally.

1894, all over again?

To avoid conflict and instead promote peace, cooperation and friendship, China and Japan must handle the Diaoyu island affair carefully, properly and competently. Indeed, the challenges faced in Sino-Japanese relations in 2012 bear an uncanny similarity to those faced by the same two neighbouring countries in the run-up to the First Sino-Japanese War of 1894–1895. It is pertinent and useful to be wary of these developments, lest history repeats itself.

First, back in 1885, Japan had decided not to act rashly after its investigations revealed that the Diaoyu islands had been already thus named by China. But, by 1887, Tokyo had started working on a five-year war plan, which was completed in 1892, with the Sino-Japanese War launched two years later. Strangely enough, shortly after Japanese coastguard vessels had harassed Chinese fishing boats in waters around the Diaoyu islands in 2010, Japan announced new national defence programme guidelines, developed a five-year military equipment plan, and shifted its military strategic focus to the southwestern islands, including Diaoyu. Generals from the Japan Ministry of Defence also proposed a variety of battle plans to cope with the Diaoyu issue and Chinese maritime strategy more generally.

Second, in 1885, warships ordered from the United Kingdom and then Germany by the Chinese Northern Ocean Fleet had been launched and rendered fully operational. Japan viewed this Northern fleet as its greatest threat, and countered by strengthening the Japanese navy. In September 2012, China's first aircraft carrier, the *Liaoning*, entered service, drawing Japan's attention and raising concerns. In recent years, Tokyo has become more wary of the development of China's military capability at sea, as well as China's sea activities. Japan is also worried that Chinese submarines would navigate larger sections of the Pacific Ocean, which go beyond Japan's monitoring scope. And therefore, while lacking financial resources, Japan nevertheless still went ahead and purchased 40 F-35s Lightning II next-generation fighter aircraft, stepped up the construction of large destroyers and submarines, and deployed the US MV-22 Osprey rotary-wing combat transport aircraft in Okinawa (Gady, 2015b; Harlan, 2011).

Third and lastly, in 1886, when four warships from the Chinese Northern Ocean Fleet visited Nagasaki, Chinese sailors on shore leave fought with Japanese police and local people using weapons and leading to various deaths, in what became known as the Nagasaki incident. Japan was forced by the Chinese fleet to release Chinese sailors who had been detained following these riots; but, as a result of this incident, Japan had been embarrassed. Anti-Qing sentiment rose nation-wide in Japan, and heralded further confrontation. Emperor Meiji mobilised the whole nation to donate money and officials were obliged to accept a pay cut of 10 per cent, so that funds could be found to build the Japanese navy in anticipation of a war against China. A strangely similar turn of events unfolded in 2010: after the Japanese coastguard showdown with Chinese fishers off the Diaoyu islands, anti-Chinese sentiment amongst the Japanese public rose; Shintaro Ishihara mobilised national donations to be able to 'purchase' the islands, and Japanese civil servants had a pay-cut of over 10 per cent.

Thus, and in summary, now that Tokyo has denied that the Diaoyu island dispute is 'shelved', and is poised to compete comprehensively with China for the control of the islands, it must make those military preparations and contingency plans necessary to cope with worst-case scenarios. Then Prime Minister Noda had declared that, if the Japan's coastguard vessels could not handle the matter alone, then Japan would dispatch its self-defence forces. Such a decision did not meet any strong objections from the Japanese Diet: a matter over which China should be highly wary. Moreover, Japan's post-1945 pacificist constitution has been amended in 2015, allowing its troops to fight abroad.

And yet, seen from a broader perspective, peace and development are in the real common interest of both China and Japan. The era of colonialism and imperialism is over. China is no longer what it was even a few decades ago; and it is difficult as well as ill-advised for Japan to resurrect its pre-World War Two militarism. World peace and development remain the mainstream of our times. China and Japan have become important trading partners to each other. As long as Japan continues to follow the road of peaceful development, Sino-Japanese relations could be stable, and even improve. However, the rise of extreme right-wing ideas and the strengthening of Japan's political right could herald a future that is full of risk for Japan; such a development would directly impact, and sour, Sino-Japanese relations.

Background to the Japanese government's 'purchase' of the islands

The Noda cabinet's decision to 'purchase the islands' has its own domestic and international settings, which mainly include the following aspects:

The domestic context

First is the domestic context. With the demise of the Soviet Union (USSR), the established post-World War Two pattern of 'cold war' confrontation between East and West disappeared. At the same time, the confrontation between conservative forces and innovative forces in Japan also shifted, and Japan's lurch towards the ideological right has become more evident. Since Junichiro Koizumi took office in 2001 as Prime Minister, the power of the political right in Japan has grown.

However, after Koizumi stepped down in 2006, these right-wing elements faced signs of decline. They sought to recover their salience and resonance with Japanese public opinion by manipulating the future development of Japanese politics, and particularly relations with China. In this context, the Diaoyu islands territorial dispute has become an 'excellent' target for the right-wing forces of Japan to instigate and secure public support for their ideas in favour of nationalism and militarisation.

In 2009, the Liberal Democratic Party of Japan lost power, and the Democratic Party took office. Since then, and until the time of writing, Japan has had three prime ministers. The Democratic Party is a party of different factions, some being doves, others being hawks. Some are friendly with China, while others are suspicious or hostile. Against a background of political turmoil, and in a context of regular elections, some Japanese politicians would sway the public to secure their votes and support. Japanese right-wing forces began to hype the Diaoyu islands affair as a tool to challenge China. A new right-wing group (Ganbare Nippon! Zenkoku Kōdō Iinkai / Hang in there, Japan! National Action Committee / 頑張れ日本！ 全国行動委員会) was set up in 2010 to stir up trouble over the Diaoyu affair, riding on, and fuelling, anti-China sentiment, while accusing the Japanese government of exhibiting weakness in Sino-Japanese relations (e.g. Nagata, 2011). In 2010, they shouted the slogan 'Down with Naoto Kan Cabinet' when the government released Zhan Qixiong, the Chinese skipper who had been detained for questioning over the alleged ramming of a Japanese coastguard vessel. During the same time, the right-wing group held anti-China rallies along every week in various locations in Japan; posted inflammatory messages over the Internet; organised trips for Japanese people to visit and land on the Diaoyu islands; and held so-called 'fishing competitions' in the same waters off the same islands.

In recent years, Japan's Ministry of Defence has had a rising influence on the country's foreign policy. In 2007, during the first administration of Prime Minister Shinzo Abe, Japan's Defence Agency was upgraded to the Ministry of Defence. As a result, the minister responsible for defence could attend cabinet meetings, and particularly meetings that involved discussions on foreign affairs. During this time, the theory of 'the China threat' became the dominant discourse in national security parlance. Thus developed a two-pronged approach: in economic terms, Japan sought to strengthen cooperation and profit from China's economic development. But, in security terms, and especially with regards to territorial issues, Japan has been increasingly wary of the increasing strength and reach of the Chinese Navy; and China is now regarded as *the* significant object of Japan's security and defence concerns. Japan seeks to extract advantage from 'the China threat' theory in order to strengthen its own military power. Thus, Japan's policy towards China is rent with contradictions. Such a distorted policy would inevitably lead to equally distorted Sino-Japanese relations featuring 'political cold and economic heat', likely to slide into a politico-economic cold.

In addition, Japan's cyber right-wing has also gained an ascendency in the digital age. Some rather extremist publications have appeared online, full of anti-China (and anti-Korea) rhetoric. It has been reported that 10 million Japanese have registered with right-wing websites, most of these being youths and working-class citizens. This trend does not bode well for Japan's ongoing and future relations with its neighbours.

This was the turbulent domestic context to 2012, the 40th anniversary of the normalisation of Sino-Japanese diplomatic relations. Shintaro Ishihara, the then controversial Governor of Tokyo, always hostile to China, was impatient. He had publically rebuked the Japanese government for not being tough enough with China and thus risking becoming the 'next Tibet'. He openly announced his decision to raise donations in order to purchase the three Diaoyu islands that were still in private hands. An online poll at the time, with 230,000 respondents, suggested that 92 per cent supported the move. Ishihara raised US$13 million in just four months. The Noda cabinet found itself powerless to stop this outburst of nationalism, and announced that the government would instead purchase the islands itself, effectively nationalising them. Allowing them to fall into the hawkish Ishihara's hands could only make matters worse. And yet, this is perhaps exactly what Ishihara wanted.

Earlier, in 2007, Japan had passed the Basic Act on Ocean Policy and then put forward a series of ocean development plans and policies. It started strengthening control over the Diaoyu islands and tried to occupy the islands in an attempt to expand its maritime rights and interests, and thus contain China's expansion in the East China Sea. The major earthquake of 11 March 2011 and the Fukushima nuclear accident have also dented Japan's development of nuclear energy. To meet Japan's energy needs, the country needs to pay more attention to the stable acquisition of hydro-carbons, apart from the development of green energy. The Diaoyu island issue must also been seen in this changing energy context.

The international context

Things have changed much in the 40-plus years since the normalisation of Sino-Japanese diplomatic relations in 1972 and the decision to shelve the Diaoyu island dispute. When speaking of the changing patterns of international relations after World War Two, the United States (US) is often considered a natural reference point. One can argue that the fundamental reason for changes in the international relations landscape lies in the rise and fall of the US as a global and regional superpower, but more significantly in the alternation of the country that is world's second largest economic power.

For example, the first notable change on the international strategic landscape after World War Two was the peaceful collapse of the Soviet Union, marking the end of the Cold War. As the second largest power at the time, the USSR was engaged in an arms race with the US, and established its own military alliance and sphere of influence, making great efforts to compete with the US on all fronts. This decision may have eventually bankrupted the USSR and sealed its demise. The US–USSR bipolarity of the 1945–1989 era then completely disintegrated, and Japan replaced the USSR to become the second largest economic power in the world. As a US ally, Japan had repeatedly stated that it shared common interests and values with the US. However, once Japan had become the world's second largest economic powerhouse, it inevitably challenged the global dominance of the US, particularly in its manufacturing industry. The US retaliated, warning Japan with a series of measures, including the appreciation of the yen. But since the 1990s, Japan's economy has faltered, with low growth levels and stubborn deflation. As a result, Japan's national gross domestic product (GDP) has fallen from 88 per cent of that of the US to around 26 per cent by 2014. For 20 or so years, Japan has been declining peacefully, and its economic heft in the world has been similarly declining: its economic activity is now about 7 per cent of the world's global output. In 2010, China replaced Japan to become the second largest economic power in the world, and was objectively viewed as the main competitor for global dominance by the US. Therefore, purely out of strategic instinct, the US has reinforced its military deployment in the Asia-Pacific region, while also consolidating the Japan–US alliance by making a commitment that the Treaty of Mutual Cooperation and Security between Japan and the US applies also to the Diaoyu islands. It is in this framework that Japan cooperates with the US in watching out for China. Since China has overtaken Japan as the world's number 2 power, Tokyo has been increasingly frustrated and uncomfortable. In this sense, the Diaoyu islands are just a political pawn on the chessboard, a strategic fulcrum in the changing state of international relations in the region and the world.

On the one hand, the US has made it quite clear that it does not hold a specific position on who has rightful sovereignty over the Diaoyu islands. On the other hand, it provides military support to Japan. Since 2010, then US Secretary of State Hillary Clinton has repeatedly stated that the Diaoyu fell within the purview of Article V of the Treaty of Mutual Cooperation and Security between Japan and the US, whereby 'each party recognizes that an armed attack against either party in the territories under the administration of Japan would be dangerous to its own peace and safety and declares that it would act to meet the common danger in accordance with its constitutional provisions and processes'. On 29 November 2012, the plenary session of the US Senate adopted a resolution that a supplementary provision be added to the National Defense Authorization Act for Fiscal Year 2013, whereby the interpretation of Article V as also covering the Diaoyu islands was confirmed. On 6 December 2012, Salvatore Angelella, Commander of US Forces in Japan, told reporters that peace and stability in the Asia-Pacific region were beneficial to all neighbouring countries including China; however, 'American troops were working closely with the Japanese to protect the country's citizens and territory' (Asahi Shimbun, 2012).

China has resolutely opposed the illegal transfer of the administrative control over the Diaoyu island chain by and from the US to Japan in 1972, pursuant to the Okinawa Reversion Treaty. Besides, Chinese law-enforcement vessels have now normalised their law-enforcement patrols in the waters off the islands; thus, the current claim by Tokyo that Japan administers the islands unilaterally is totally false. In addition, even from the position of the US, administration does not mean sovereignty, and the US neither recognises Japanese sovereignty over the islands, nor admits the islands as Japanese territory. Therefore, why should the

Diaoyu islands feature at all within the scope of the Treaty of Mutual Cooperation and Security between Japan and the US?

The Noda cabinet 'purchase' of the remaining Diaoyu islands still in private hands was, in essence, an attempt to achieve long-term stable and secure administration by Tokyo over the island chain, and eventually to permanently occupy these islands. This is a trick, meant to turn an illegal activity into a legal one. Should the government of a country maintain peaceful, stable, effective and continuous occupation and administration over the territory of another country with no effective opposition to such acts over a long period of time, then that government has a strong basis to claim that place as its sovereign territory in accordance with international law. One can conclude that the ultimate purpose of the Noda cabinet decision to 'nationalise' the islands was one step towards such a regularisation. However, this is really an impossible mission.

On 13 May 2012, during Japanese Prime Minister Noda's visit to China, Premier Wen Jiabao stated Beijing's position on the Diaoyu island affair unequivocally. But when Noda returned to Japan, he decided, just five days later, to discuss the 'purchase' of the islands. On 9 September 2012, during the informal meeting of the Asia-Pacific Economic Cooperation (APEC) in Vladivostok, Russia, President Hu Jintao also told Noda in person that Sino-Japanese relations were facing a serious situation due to developments around the Diaoyu islands, and that Beijing was resolutely opposed to the attempts being made by Japan to 'purchase' the islands. China would consider any such act as illegal, invalid and provocative; the Japanese government should be fully aware of the seriousness of the situation, and should stop making the wrong decisions. Instead, Japan had better work with China to safeguard the overall development of Sino-Japanese relations. To no avail: regardless of Beijing's repeated protestations and warnings, the Japanese government doggedly went ahead and announced the purchase of Diaoyu Dao/Uotsuri-shima and its two affiliated islets of Bei Xiaodao/Kita-Kojima and Nan Xiaodao/Minami-Kojima, thus 'nationalising' them.

The significance of China's counter-measures

On the same day that the Japanese government announced that it would 'purchase' the islands, the Chinese government announced that it would by no means put up with such a blatant encroachment of its own territory. Beijing announced the territorial base points and baselines in accordance with the Law of the People's Republic of China on the Territorial Sea and the Contiguous Zone. On the same day, Chinese Foreign Minister Yang Jiechi urgently summoned the Japanese Ambassador to China, Uichiro Niwa, and made serious representations and strong protestations against the Japanese government's illegal 'purchase' of the three Diaoyu islands. The Chinese Ambassador to Japan, Cheng Yonghua, also made serious representations to Japan's Ministry of Foreign Affairs, and also submitted a formal note of protest. All the commemorative activities planned to celebrate the 40th anniversary of the normalisation of Sino-Japanese relations on 27 September were summarily called off.

The main significance and effects of China's counter-measures in the wake of the 'purchase' decision are two.

First is the normalisation of China's patrol and law-enforcement capability around the Diaoyu islands, which has broken Japan's earlier unilateral control over, and presence around, these same islands. Since Japan's original decision to 'shelve the dispute' has changed, the groundwork to return to a situation of 'shelving' as normalcy now no longer exists. A unilateral adherence to 'shelving' by China is not realistic, and China should move and adapt to the changing modalities of the Diaoyu affair. As to the territorial sovereignty of the islands, the one thing that China can certainly do is to continue the tit-for-tat confrontation to safeguard its claims for sovereignty over the Diaoyu group. On 14 September 2012, Chinese marine surveillance ships 15, 26, 27, 50, 51 and 56 entered waters within 12 nautical miles off the Diaoyu islands. This was the first time that Chinese ships patrolled in these waters since the Chinese government announced the territorial baselines of Diaoyu Dao/Uotsuri-shima and its affiliated islands. Moreover, the size of the surveillance fleet was unprecedented. Even during the national holiday season (1–8 October 2012), the fleets continued their patrols every day, an act indicative that Beijing claimed and exercised jurisdiction over the waters surrounding the Diaoyu islands, and that Beijing was resolute in the regularisation of its patrolling activities. What the Chinese government has done is to totally break up Japan's so-called 'positive prescription' and 'stable and secure administration', while also demonstrating for all to see that China was exercising its sovereignty over the islands.

Second is that, through its various efforts, China has decidedly taken the upper hand in the Diaoyu islands affair. On 25 September 2012, the State Council Information Office of the People's Republic of China published a white paper entitled *Diaoyu Dao: An Inherent Territory of China*, which gives a complete and systematic description of the reasons behind China's sovereignty claims over the islands (State Council of the PRC, 2012). In this scenario, the Japanese government has been caught in a dilemma: keeping silent in the face of China's counter-measures would be interpreted as weakness; while any escalation was risky and dangerous as well as ridiculous, since the Japanese official line is that really there is no dispute over the islands.

Making lame and shameless arguments

The more Japan tries to hide, the more it is exposed. On 10 October 2012, at a press conference, Foreign Minister Koichiro Genba of the Noda Cabinet could only quote the publicity material used by Japanese right-wing forces, claiming that China used to admit that the Diaoyu islands were Japanese territory. This material included a world atlas published in China in 1958 and republished in 1960, in which Japan's Okinawa was marked as 'the Senkaku islands', as well as a certificate of appreciation to Japan from the Chinese consul at Nagasaki in 1920, which mentions 'the Senkaku islands'.

According to sources from the Japanese Foreign Ministry, the Japanese government views the absence of any protest by China against Japan for 'occupying' the

Diaoyu islands before 1970 as the biggest contradiction in China's claim for sovereignty. Based upon an article published in the *People's Daily* on 8 January 1953, Japan argues that China used to recognise these islands as Japanese territory, and declared so publicly to the international community.

The above practices show that the stance of the Japanese government on the Diaoyu islands is baseless, both in terms of historic fact and legal grounds. The Japanese government has only to hang on to the rhetoric of the Japanese right-wing forces to make lame and shameless arguments, as will be explained below.

First, in 1920, the Chinese consul in Nagasaki awarded a 'certificate of appreciation' to the Japanese for rescuing some Chinese fishers. This certificate referred to the 'Senkaku Islands, Yaeyama District, Okinawa Prefecture, the Empire of Japan'. However, when one considers that Taiwan was then ruled by the Japanese, such 'evidence' was in fact baseless with regards to sovereignty claims.

Second, on the world atlas published in 1958, the Diaoyu islands belonged to Okinawa. However, the title page of this publication specified that part of China's borders had been drawn based on maps submitted before World War Two. In the *World Map by Countries*, published by China in 1956, Okinawa did not include the Diaoyu islands.

Third, an article that appeared in the Chinese *People's Daily* on 8 January 1953 stated that the Ryukyu Islands included the Diaoyu/Senkaku islands. After an investigation, the article turns out to be material that had been translated and edited from Japanese sources without attribution: evidence of this is that the word 'Kadena' in the article was translated into 'Katana' (transliteration) from the Japanese pronunciation. And so, this could hardly represent the Chinese government's official position on the Diaoyu islands.

Finally, Tokyo often argues that China started to claim sovereignty over the Diaoyu island chain only from the latter half of 1970, after oil deposits had been discovered in the surrounding seas. This is a complete lie. Already on 18 May 1970, the *People's Daily* had already published an article that condemned the intentions of Japan Prime Minister Eissaku Sato's cabinet to use Huangwei Islet/Kuba-shima and Chiwei Islet/Taishō-tō as shooting ranges so as to prove Japan's sovereignty over the islands. After the end of World War Two, the Japanese Ministry of Foreign Affairs did not make any official sovereignty claim over the Diaoyu islands before March 1972.

Prime Minister Yoshihiko Noda's decision to 'purchase' the islands damaged Sino-Japanese relations, resulting in frostier economic and political relations. Japanese business suffered a major blow. With the Noda cabinet decision to 'purchase' the Diaoyu islands, some 100,000 Chinese tourists planning to visit Japan cancelled their trips. Various flights were annulled; other flights were suspended. The production volume of Japan's six major automobile manufacturers decreased by over 8 per cent in the wake of the decision. The Noda 'purchase' caused more economic damage to Japan than the Fukushima nuclear-power plant accident.

Prospects for Sino-Japanese relations

China and Japan should confirm that peace and friendship are the foundations of Sino-Japanese relations, defuse the dangerous stand-off they have found themselves in, and avoid unnecessary conflict over the Diaoyu islands.

Now, there are two totally different policy options concerning these islands: zero-sum options and positive-sum (win–win) options. The future fate and trajectory of Sino-Japanese relations may very well hinge on the critical choices that each side makes, including on the sensitive issue of the Diaoyu islands. With the tense political climate that unfolded in 2012, people of insight and those in a position to influence policy from both China and Japan should make every effort to cherish and protect Sino-Japanese relations. Rather than condone or undertake actions that damage Sino-Japanese relations, Chinese and Japanese people and their politicians should work together to form a united front in order to uphold the right path to peace and progress.

Even though both China and Japan do not want to fight each other, China believes that it can only ensure and guarantee its own peace and security by enhancing its national defence capabilities. In response to this, Japan is likely to take measures to further strengthen its coastguard, air reconnaissance, maritime policing and defence force deployment in and around the Diaoyu islands. Japan will further enhance its cooperation with the US, as well as with other countries, so as to show and remind China of the Japan–US alliance and other allies in the region, many of whom may be nervous over China's policies in the South China Sea. This cooperation is now rendered easier with the 2015 constitutional amendment that approves the right to collective self-defence. Japan has also struck a separate fisheries agreement with Taiwan, seeking to divide the Chinese position on this sensitive matter. However, such an escalation may have the workings of a self-fulfilling prophecy, trapping the players into a vicious spiral, and evolving into a comprehensive confrontation between Beijing and Tokyo, even involving Washington, DC and thus leading to the deterioration of China–US relations as well. The arrest of the Chinese skipper in 2010, the landing of Japanese nationalists on the disputed islands, the Noda cabinet decision to 'purchase' three islands in 2012, and the persistent Japanese government denial that there is even a dispute – all these factors contribute to this downward dangerous and vicious spiral. So much is clear and self-evident.

Towards a virtuous spiral?

This is why China and Japan should make every effort to escape this vicious road map and switch to an alternative, virtuous road map; starting a dialogue over the Diaoyu islands may be the welcome start to such a switch. Maritime mutual confidence-building measures and crisis-management mechanisms need to be established. Fishing rights in the waters around the Diaoyu islands for those on both sides across the Taiwan Strait need to be secured through a re-negotiation of the China–Japan Fisheries Agreement. If the official channel is not so convenient, given so much bad blood between the two countries, then non-governmental experts and other 'lower-level' personnel can come into play. In addition, China and Japan can jointly investigate and study legal and other related issues concerning the ownership of the islands through scientific, academic and cultural exchanges. Otherwise, China and Japan may also simply agree to disagree and decide, once again, to shelve the issue, returning to the *status quo ante* of 1972.

Meanwhile, whether this shelving happens or not, nothing is really preventing both sides from taking steps that reflect a mutually beneficial, 'win–win' approach. These may include an agreement to develop the Diaoyu islands, and their surrounding waters, in economic and commercial ways, completely avoiding and shelving the controversial sovereignty issue. Admittedly, the joint exploration for hydro-carbon and other energy deposits in deep offshore waters is not such an attractive proposition as long as the price of oil on the international markets is so inexpensive. However, joint sightseeing tourist activities can be organised, in collaboration with Taiwan and Okinawa.

Conclusion

The year 2018 will mark the 40th anniversary of the signing of the Sino-Japanese Treaty of Peace and Friendship. The China–Japan relationship should, and can, return to the correct spirit as established in 1978 by this landmark treaty. Based on the content and spirit of this treaty, China and Japan should first reiterate that they will pursue peaceful means to settle all disputes, and will refrain from the use of force, or the threat thereof, in their bilateral dealings. The Diaoyu affair would be included in the remit of such a declaration.

China and Japan can transform challenges into opportunities, thus easing tensions in the Taiwan Strait, while embarking on joint development initiatives, including offshore sightseeing tourism involving Okinawa and Taiwan. Such actions are likely to prove highly significant to the healthy development of the China–Japan–US triangular relationship, as well as to the formation of a new pattern of international relations in the region. Instead of coastguard personnel, vessels and other air/sea craft from China and Japan eyeing each other closely in a dangerous game around a group of small islands, would it not be better to have fishers from China and Japan engaged in sustainable fishing, or tourists from China and Japan enjoying a visit to these islands while on a trip that would have included mainland China, Taiwan and Okinawa as stops?

Epilogue II

Proposing solutions to the Senkaku/ Diaoyu affair

Akihiro Iwashita

Introduction: four-player versus two-state solution

As with the Kurile Islands and Takeshima/Dokdo, the Senkaku/Diaoyu islands represent one of the pending territorial disputes of Japan with its neighbours Russia, Korea and China respectively. In the case of the Senkaku/Diaoyu islands affair, I concur with Professor Baldacchino who, in this timely book, positions the matter squarely in a four-player context, rather than as a stark, two-state showdown. It is in the critical roles played by Taiwan and Okinawa that may lurk the opportunities to deftly shift the matter from a bilateral nationalist confrontation into one that involves regional players, each with very different – and less confrontational – historical memories than their putative patron states. Therein may lie an eventual solution to the conundrum lingering in the East China Sea. Various avenues for joint cooperation and development, including the various imaginative 'solution protocols' outlined in this book, and possibly others as well, may become much more realistic, feasible and sustainable when, and if, Taipei and Naha join Beijing and Tokyo in the equation.

Reviewing the main claims

Let us start with a look at the main claims. For Japan, it was the Meiji government who investigated the local situation, confirming the lack of control over the Senkaku islands by the Qing dynasty in China, and started to administer them by right of first occupation of a *terra nullius*, with the formal cabinet decision to do so being taken in January 1895. The occupation of the islands had nothing to do with the Sino-Japanese war; and so the islands were never included in the territory that Japan gave up as a result of signing the San Francisco Peace Treaty; the islands were controlled by the US after World War Two and were returned to Japan, along with the rest of Okinawa, in 1972. China's claims to the islands began in the early 1970s, particularly after the publication of reports confirming the existence of oil and gas resources in the sea around the islands. Prior to this, from 1895 (when the First Sino-Japanese War ended) to the 1970s, China had never once claimed the islands from Japan. The Communist Party newspaper *The People's Daily* had previously stated that the islands had belonged to the Ryukyus, and hence were now a part of Japan (Ministry of Foreign Affairs, Japan, 2015).

China rebuffs Japan's claims; it argues that the Diaoyu islands are mentioned in Chinese historical records dating back to the Ming dynasty and never belonged to the Ryukyu kingdom. Japan deprived China of these islands by force during the First Sino-Japanese War: the Diaoyu islands were included in the islands around Taiwan that Japan seized in the wake of that conflict. They are therefore part of a history of Japanese aggression towards China. Nor was China party to the San Francisco Peace Treaty.

The dangers of historicisation

Both Japan and China have 'rediscovered' the Senkaku/Diaoyu island chain as part of their national history recently. From a jurisdictional point of view, the critical year is 1895; and Japan would enjoy the advantage, given the long, 75-year silence from China over its claims to these islands. And yet, the islands have been catapulted to the status of an 'inherent and integral territory' of either country in its respective political rhetoric. When such insignificant pieces of land are manipulated politically, and their status transformed into issues of history, territorial disputes as in the Senkaku/Diaoyu case grow into uncontrollable monsters. Which is why the case of these islands should not be allowed to become a historical, and hence a national, issue.

Meanwhile, a United States far from Northeast Asia is politically able to be close to both China and Japan. As a result, it is essential for Tokyo to constantly nurture its ties with Washington through an affirmation of their critical joint interests. If a 'common enemy' is put in place, then no one will question the value of the alliance, and indeed will find some solace in its existence.

The April 2012 announcement by Shintaro Ishihara, then Governor of Tokyo, that 'Tokyo would buy the Senkakus' made exactly this point. In order to sensationalise this decision for the media and the general public, 11.4 billion yen were collected and Japanese nationalist fervour raised to fever pitch. This occurred, even though the possession of the islands is a matter separate from issues of sovereignty, and so the intended purchase by Tokyo would have done nothing to 'protect' them. However, a visibly perturbed Japanese government decided that the state would buy the Senkakus instead, effectively nationalising the three remaining islands still in private hands, for a reported 2 billion yen. Tokyo presumed that Beijing would understand this action, and Prime Minister Noda had warned then Chinese Premier Hu Jintao beforehand of this move; but the result was exactly the reverse. Premier Hu strongly criticised Noda, and China was again rocked by anti-Japanese demonstrations. Conventions and events were cancelled or postponed; economic relations cooled; Japanese investment in China targeted and vandalised; and the number of Chinese tourists visiting Japan collapsed. A car carrying the Japanese ambassador in the Chinese capital was attacked by a man who ripped the Japanese flag off (Foster, 2012).

In this scenario, criticism from Japanese sources of the potential relocation (let alone closure) of US bases in Okinawa became more difficult to justify because of the 'threat of China'. The stationing of Osprey helicopters on Okinawa, as well as of some 500 troops from the Japan Self-Defence Force on Ishigaki (*Japan Times*, 2015), was now justified with reference to the 'defence' of the Senkakus and of Japan itself.

Tokyo completely miscalculated Beijing's reaction to the 'purchase': China went immediately on the offensive. By continuously violating the air space and territorial waters around the Senkaku/Diaoyu islands, China would claim this action as evidence of its own jurisdictional authority, both domestically and externally. China also started asserting the historical nature of its claim to the islands and the surrounding waters, seeking to overturn the results of World War Two when, in spite of the Cairo Declaration commitments, the allied powers had brazenly *not* returned the Diaoyu islands, along with Taiwan, to China. This rhetorical formula highlights Japan's continuing imperialistic adventures as well as the implicit connivance of the United States in this behaviour. In so doing, China also links its claims to the Diaoyu islands with those to Taiwan. Since Taiwan was separated from China and given to Japan with the Treaty of Shimonoseki that brought the First Sino-Japanese War to an end in 1895, Beijing is able to argue that Taiwan, like the Diaoyu islands, is part of China. Such a fantastic story would come to seem real were China to actually gain even temporary control of the islands.

One party that is not likely to buy this argument is Taiwan itself. First, if Taiwan is being linked to the Senkaku/Diaoyu islands and claimed as Chinese, then Taipei cannot accept a 'restoration' of the Senkaku/Diaoyu islands that would imply the similar 'restoration' of Taiwan. Second, the Taiwanese do not share the memories of the atrocities committed in China during the Second Sino-Japanese War when thinking about World War Two. Instead, the lingering image in Taiwan is of the damage inflicted by the Chinese mainland with the start of Kuomintang rule after 1949, and after the Japanese had pulled out. Third, since Tokyo does not recognise Taiwan as a state, there is still, opportunistically, no 'national' border between Taiwan with Japan.

For a plurality of regional actors and voices

In any event, it would be advisable to urgently prevent China's efforts to historicise the Senkaku/Diaoyu issue. This is because it is seeking Japanese recognition for what is really only a territorial issue, and also because the affair becomes memorialised domestically, while dragging in neighbouring states with similar historical qualms (such as Korea). Similarly, it is equally critical for Tokyo to avoid creating its own national story that casts China squarely as the villain of the piece for the sake of the Japan–US alliance. The de-historicisation of the issue is a precondition for the resolution of the matter. Instead of lingering historical bitterness, a plurality of strong regional actors can take over and dominate the discourse over the contested island chain.

Frank opinions and conversations about modern history need to continue. Japan must listen to China's claims over past Japanese imperial activities. Details about episodes, such as the Ryukyuan tribute missions to the Qing court, the Japanese 'incorporation' of Ryukyu, and Tokyo's offer to give China the Ryukyu islands of Miyako and Yaeyama and everything to the south of them to Taiwan, before the First Sino-Japanese War, in exchange for commercial rights in China – so much for reference to inherent and integral territory! – should be discussed freely, and involving local historians from Yaeyama and Okinawa (Drifte, 2014). One should know of the histories of the relations of Yonaguni, Ishigaki and Miyako with the Senkaku/Diaoyu. Also involved should be historians from Taiwan, not just from Taipei but from the east coast communities of Hualien, Keeling and Suao. That the waters to the south of the Senkaku/Diaoyu islands have been previously used by both Yaeyama fishers and the Taiwanese should be remembered and acknowledged. The long history of connectivities, migrations and mobilities will be privileged from such an exchange, eclipsing the stark and divisive political oratory of borders, sovereignty and territoriality.

Moreover, 'border conflicts' can look very different from a local perspective. While national politicians, army officers and other hawks engage in sabre rattling over the presumed threat to sovereignty and the imagined community of the nation, the local population on either side of the political divide may harbour uncannily common interests. The local residents may not be interested in the full, or even partial, 'recovery' of disputed territory within the national fold. Instead, they may yearn for flexible arrangements and access to resources – such as fishing grounds – and the stabilisation and normalisation of area relations which lie at the basis of other critical activities such as the attraction of external direct investment, tourism development as well as population growth.

What else, then, can be done to help resolve the Senkaku/Diaoyu matter? For the Japanese government, the response would probably be that there is no dispute to resolve in the first place, thus sniffing out any promising conversation. But, the recent upsurge in tension cannot just be neglected. Rather than falling into the current 'historical trap' – talking about territorial matters from a 'zero-sum' statist perspective – ideas should be sought to resolve the affair through the effective co-management of the islands and the contiguous maritime area, with opportunities for positive sum gains. With the advantage of hindsight, it appears that the Noda Cabinet decision to nationalise the Senkakus was a mistake; even though most Japanese would have sided with their government in supporting the decision in principle, if anything to deter then Tokyo Governor Ishihara from making the purchase and whatever else might have followed. Moreover, the US administration has been obliged to clarify that any attack on the Senkakus would be seen as triggering the Japan–US military alliance into action. However, nationalisation has *not* improved control over the islands; if anything, Beijing's response to this decision has made Japanese control *less* effective. It would therefore be quite appropriate to better explore actual control options and how to securely manage the islands in a manner conducive to regional security, peace and economic prosperity.

Conclusion: the return of the albatross?

One appealing suggestion that places importance on the sea as well as the Senkaku/Diaoyu islands proper, departs from the perspective of protecting the area's natural resources, long exploited by humans. One idea would be to set up some sort of 'Albatross Foundation', entrust the management of the islands in its hands, and 'privatise' it in that way, away from the meddlesome aspirations of politicians. This 'Albatross Foundation' would support other life forms on land, as well as fish stocks in the surrounding sea. Tatsushiro Koga, the man who owned the bonito processing plant on Uotsuri-shima/Diaoyu Dao, previously hunted albatross to extinction for their feathers on these islands. However, surveys by Hiroki Hasegawa of Toho University in 1991–1992 and again in 2001–2002 found indications that the albatross had started to return to breed on the islands, particularly on Kita-Kojima/Bei Xiaodao and Minami-Kojima/Nam Xiaodao. Said to have become extinct in 1949, the birds are slowly returning and recolonising their natural habitat, now without human population or visitation. There may be other rare and/or endemic species on the islands, and culling the goats that are currently running wild there would probably be a wise move. It would also help if links were formed with international organisations to help protect and preserve the islands' natural habitat. Of course, researchers from China and Japan, and inclusive of Taiwan and Okinawa, and not just the West, should be involved. The involvement of Tokyo's metropolitan government would also be a bonus, since this has recently undertaken a cull of goats on the Ogasawara islands; while Tori island, in the Izu island chain under Tokyo's jurisdiction, is the largest albatross breeding ground in the world. Indeed, the money that Tokyo collected when Ishihara was hoping to buy the islands in 2012 could now be used to support this plan.

Such steps would increase mutual trust, promote stability and peace in an otherwise tense region of the globe, while establishing a new Japan–China–Taiwan–Okinawa quadrilateral consensus not to 'touch' the sovereignty issue.

Islands continue to feature prominently in experiments dealing with creative governance. The recent '50–50' division of Heixiazi/Bolshoi Ussuriiskii island between China and Russia (see Chapter 7) creates an interesting precedent in East Asia, and one that has also involved Beijing. Might not the East China Sea become the location for another example of successful co-management of disputed territory?

References

ABC (2013, 5 December). East Timor spying case: PM Xanana Gusmão calls for Australia to explain itself over ASIO raids. *Australian Broadcasting Corporation.* Retrieved from www.abc.net.au/news/2013-12-04/east-timor-pm-shocked-by-asio-raids-over-spying-case/5135146

Abulafia, D. (2011). *The great sea: A human history of the Mediterranean.* Oxford: Oxford University Press.

Ackrén, M. and Lindström, B, (2012). Autonomy development, irredentism and secessionism in a Nordic context. *Commonwealth & Comparative Politics, 50*(4), 494–511.

Act on the Autonomy of Åland (2004). Unofficial translation, the original text in Swedish and Finnish, published by the Åland Parliament and the Government of Åland, Mariehamn.

Adviesgroep Waddenzeebeleid (2004). *Ruimte voor de Wadden: Eindrapport Adviesgroep Waddenzeebeleid.* The Hague: VROM en EZ (in Dutch).

Agamben, G. (2005). *State of exception.* Translated by Kevin Attell. Chicago, IL: University of Chicago Press.

Agreement between Japan and Korea (1974, 30 January). *Agreement between Japan and the Republic of Korea concerning joint development of the southern part of the continental shelf adjacent to the two countries.* Entered into force 22 June 1978. Retrieved from www.un.org/Depts/los/LEGISLATIONANDTREATIES/PDFFILES/TREATIES/jakor1974south.tif

Agreement between Norway and Iceland (1980, 28 May). *Agreement between Norway and Iceland on fishery and continental shelf questions.* Entered into force 13 June 1980. Retrieved from www.un.org/Depts/los/LEGISLATIONANDTREATIES/PDFFILES/TREATIES/isl-nor1980fcs.pdf

Agreement regarding Iceland and Jan Mayen (1981, 22 October). *Agreement on the Continental Shelf between Iceland and Jan Mayen.* Entered into force 2 June 1982. Retrieved from www.un.org/Depts/los/LEGISLATIONANDTREATIES/PDFFILES/TREATIES/ISLNOR1981CS.PDF

Ahlström, C. (2004). *Demilitarised and neutralised territories in Europe.* Mariehamn: The Åland Islands Peace Institute.

Al-Alkim, H. H. (2002). The islands question. In G. Sick and L. Potter, L. (eds), *Security in the Persian Gulf: Origins, obstacles, and the search for consensus* (pp. 155–170). New York: Palgrave Macmillan.

Åland Culture Foundation (2010). *International documents 1856–2009 about the special status of the Åland Islands* (2nd edn). Mariehamn: Åland Culture Foundation.

Aldrich, R. and Connell, J. (1998). *The last colonies*. Cambridge: Cambridge University Press.

Alexander, W. J. (1999). Science, history and the Kasikili Island dispute. *South African Journal of Science, 95*(8), 321–324.

Ali, S. (2010). *Dubai: Gilded cage*. New Haven, CT: Yale University Press.

Aliberti, M. (2015). China, the moon, and the world. In *When China goes to the Moon* (pp. 185–252). New York: Springer International.

Allison, G. T. and Zelikow, P. (1999). *Essence of decision: Explaining the Cuban missile crisis*. New York: Longman.

Anderson, B. (1983/2006). *Imagined communities: Reflections on the origin and spread of nationalism*. London: Verso.

Anderson, E. W. (2000). *Global geopolitical flashpoints: An atlas of conflict*. London: The Stationery Office.

Anderson, E. W. (2003). *International boundaries: A geopolitical atlas*. New York: Routledge.

Anthony, P. A. (1998). Popular Catholicism and social change in St Lucia. *Social Compass, 45*(4), 555–574.

AOSIS (2015). *Alliance of Small Island States*. Retrieved from http://aosis.org/

Aosis.org (2015, 12 December). Closing Statement Paris Agreement. Retrieved from http://aosis.org/closing-statement-paris-agreement/

Arlov, T. B. (2011). *Den rette mann: Historien om Sysselmannen på Svalbard*. Trondheim: Tapir.

Arndt, O., Dirks, H. and Koch, T. (2004). *The Wadden Sea region: A socio-economic analysis*. Bremen: Prognos AG.

Asahi Shimbun (2012, 26 December). Normalisation of relations: China claims it agreed with Japan to shelve the dispute in 1972, Japan denies. Retrieved from http://ajw.asahi.com/article/special/Senkaku_History/AJ201212260103

Asahi Shimbun (2013, 11 May). Japan–Taiwan fishing pact takes effect, but rough waters lie ahead. Retrieved from http://ajw.asahi.com/article/behind_news/politics/AJ201305110065

Atauz, A. D. (2008). *Eight thousand years of Maltese maritime history: Trade, piracy and naval warfare in the Central Mediterranean*. Gainesville, FL: University Press of Florida.

Åtland, K. and Pedersen, T. (2009). Russia's Svalbard policies: Securitisation still in fashion? Paper presented at the 50th Annual Convention of the International Studies Association, New York, February.

Auburn, F. M. (1982). *Antarctic law and politics*. Bloomington, IN: Indiana University Press.

Austin, G. (1998). *China's ocean frontier: International law, military force and national development*. Canberra ACT: Allen & Unwin.

Ayumi, T. (2015). Adhering, distancing or waffling? Understanding a new dilemma in the U.S.–Japan alliance. *Georgetown Journal of Asian Affairs*, Spring/Summer, 67–97.

Ayusawa, S. (1964). Geography and Japanese knowledge of world geography. *Monumenta Nipponica, 19*(3–4), 275–294

Badejo, F. A. (2008). *Claude: A portrait of power*. Philipsburg, Sint Maarten: House of Nehesi.

Bahcheli, T., Bartmann, B. and Srebrnik, H. F. (eds) (2004). *De facto states: The quest for sovereignty*. New York: Routledge.

Baker, C. W. (2015). The dismantlement of the Netherlands Antilles: how autonomy is more viable than independence. *Caribbean Quilt*, *3*(1), 142–153.

Baldacchino, G. (1993). Bursting the bubble: the pseudo-development strategies of microstates. *Development and Change*, *24*(1), 29–51.

Baldacchino, G. (2006). Innovative development strategies from non-sovereign island jurisdictions? A global review of economic policy and governance practices. *World Development*, *34*(5), 852–867.

Baldacchino, G. (2007). Islands as novelty sites. *Geographical Review*, *97*(2), 165–174.

Baldacchino, G. (2010). *Island enclaves: Offshoring strategies, creative governance, and subnational island jurisdictions*. Montreal and Kingston: McGill-Queen's University Press.

Baldacchino, G. (2012a). Governmentality is all the rage: the strategy games of small jurisdictions. *The Round Table: Commonwealth Journal of International Affairs*, *101*(3), 235–251.

Baldacchino, G. (2012b). Islands and despots. *Commonwealth & Comparative Politics*, *50*(1), 103–120.

Baldacchino, G. (ed.) (2013). *The political economy of divided islands: Unified geographies, multiple polities*. New York: Palgrave Macmillan.

Baldacchino, G. (2015). Capital and port cities on small islands sallying forth beyond their walls: a Mediterranean exercise. *Journal of Mediterranean Studies*, *23*(2), 137–151.

Baldacchino, G. and Hepburn, E. (2012). *Independence, nationalism and subnational island jurisdictions*. London: Routledge.

Baldacchino, G. and Mellor, C. S. (2015). Tuvalu: entrepreneurship and the Dot TV phenomenon. In G. Baldacchino (ed.), *Entrepreneurship in small island states and territories* (pp. 268–281). New York: Routledge.

Baldacchino, G. and Milne, D. (2006). Exploring sub-national island jurisdictions. *The Round Table: Commonwealth Journal of International Affairs*, *95*(386), 487–502.

Baldacchino, G. and Tsai, H. M. (2014). Contested enclave metageographies: the offshore islands of Taiwan. *Political Geography*, *40*(1), 13–24.

Bandow, D. (2014, 27 December). U.S. should close bases on Okinawa: bring troops home and let Japanese defend Japan. *Huffington Post*. Retrieved from www.huffingtonpost.com/doug-bandow/us-should-close-bases-on-okinawa_b_6384110.html

Banyan (2014, 6 December). The world is Xi's oyster. *The Economist*. Retrieved from www.economist.com/news/china/21635516-confident-chinese-leader-sets-out-his-foreign-policy-store-it-not-wholly-comforting-world

Barros, J. (1968). *The Åland island question: Its settlement by the League of Nations*. New Haven, CT: Yale University Press.

Bartmann, B. (2002). *Saltwater frontiers: Jurisdiction as a resource for small islands*. Charlottetown: Institute of Island Studies, University of Prince Edward Island.

Bazelmans, J., Meier, D., Nieuwhof, A., Spek, T. and Vos, P., (2012). Understanding the cultural historical value of the Wadden Sea region: the co-evolution of environment and society in the Wadden Sea area in the Holocene up until early modern times (11,700 BC–1800 AD). *Ocean and Coastal Management*, *68*(1), 114–126.

BBC News (2011, 21 December). South American states ban Falklands vessels from ports. Retrieved from www.bbc.com/news/uk-16280613

BBC News (2012a, 17 April). Tokyo to buy disputed islands, says Governor Ishihara. Retrieved from www.bbc.com/news/world-asia-17747934

BBC News (2012b, 10 September). Japan confirms disputed islands purchase plan. Retrieved from www.bbc.com/news/world-asia-19540469

BBC News (2013, 23 November). China establishes 'air-defence zone' over East China Sea. Retrieved from www.bbc.com/news/world-asia-25062525

BBC News (2015, 16 July). Japan's lower house approves change to self-defence law. Retrieved from www.bbc.com/news/world-asia-33546465

Beagle Channel Cartography (2016). Retrieved from https://en.wikipedia.org/wiki/Beagle_conflict#/media/File:Resume.arg.cart.en.png

Beckman, R., Schofield, C., Townsend-Gault, I., Davenport, T. and Bernard, L. (2013). Factors conducive to joint development in Asia: lessons learned for the South China Sea. In R. Beckman, I. Townsend-Gault, C. Schofield, T. Davenport and L. Bernard (eds), *Beyond territorial disputes in the South China Sea: Legal frameworks for the joint development of hydrocarbon resources* (pp. 291–311). Cheltenham: Edward Elgar.

Beijing Guide (2015). Summer Palace (Yiheyuan). Retrieved from www.china.org.cn/english/MATERIAL/30960.htm

Bender, J. (2015, 5 May). China is making moves in Antarctica. *Business Insider*. Retrieved from www.businessinsider.com/china-is-making-moves-in-antarctica-2015-5#ixzz3jCsxGxhi

Blank, S. (2011, 29 July). Revising the border: China's inroads into Tajikistan. *China Brief 11*(14). Retrieved from www.jamestown.org/single/?tx_ttnews[tt_news]=38251&no_cache=1#.Vd3Ld5cYM90

Blatter, J. K. (2001). Lessons from Lake Constance: ideas, institutions, and advocacy coalitions. In J. Blatter and H. M. Ingram (eds), *Reflections on water: New approaches to transboundary conflict and cooperation* (pp. 89–122). Cambridge, MA: MIT Press.

Blatter, J. K. (2004). From 'spaces of place' to 'spaces of flows'? Territorial and functional governance in cross-border regions in Europe and North America. *International Journal of Urban & Regional Research 28*(3), 530–549.

Blatter, J. K. and Ingram, H. M. (2001). *Reflections on water: New approaches to transboundary conflicts and cooperation*. Cambridge, MA: MIT Press.

Blum, J. (2013a, 22 July). 保卫钓鱼岛！（光荣使命）[To defend the Diaoyu Islands! Glorious mission] Retrieved from www.youtube.com/watch?v=LdpSgEn73ms

Blum, J. (2013b, 22 July). Chinese online game lets players 'reclaim and defend the Diaoyu Islands'. *South China Morning Post*. Retrieved from www.scmp.com/lifestyle/technology/article/1288462/chinese-online-game-lets-players-reclaim-and-defend-diaoyu

Boere, G. C. and Piersma, T. (2012). Flyway protection and the predicament of our migrant birds: a critical look at international conservation policies and the Dutch Wadden Sea. *Ocean and Coastal Management, 68*(1), 157–168.

Bong, Y. D. (2013). Built to last: The Dokdo territorial controversy. The baseline conditions in domestic politics and international security of Japan and South Korea. *Memory Studies, 6*(2), 191–203.

Bradford, E. (1961/1999). *The Great Siege: Malta 1565*. Ware: Wordsworth Editions.

Bradsher, K. (2010, 22 September). Amid tension, China blocks vital exports to Japan. *New York Times*. Retrieved from www.nytimes.com/2010/09/23/business/global/23rare.html?_r=0

Brant, R. W. and Maycock, W. (eds) (1911). General instructions to the British High Commissioner with regard to the application of the convention of October 20, 1906. In *British and foreign state papers: 1906–1907*. London: HMSO.

Braudel, F. (1958). Histoire et sciences sociales: la longue durée. *Annales. Histoire, Sciences Sociales, 13*(4), 725–753. Paris: École des Hautes Études en Sciences Sociales (in French).

Braun, B. (2000). Producing vertical territory: geology and governmentality in late Victorian Canada. *Cultural Geographies*, *7*(1), 7–46.

Broches, A. (1987). Awards rendered pursuant to the ICSID Convention: binding force, finality, recognition, enforcement, execution. *ICSID Review 2*(2), 287–334. Retrieved from http://icsidreview.oxfordjournals.org/content/2/2/287.full.pdf+html

Brook, T. and Wakabayashi, B. T. (2000). *Opium regimes: China, Britain, and Japan, 1839–1952*. Los Angeles: University of California Press.

Brunet-Jailly, E. (2015). *Border disputes: A global encyclopedia*. 3 vols. Santa Barbara, CA: ABC-CLIO.

Brunton, R. (1997). Black and white: the cannibalised academy. *Review-Institute of Public Affairs*, *49*(4), 23–24. Retrieved from https://ipa.org.au/library/publication/1226023398_document_1211162509_document_review1997_vol.49no.4.pdf

Buck, S. L. (2012). *The global commons: An introduction*. Washington, DC: Island Press.

Buerk, R. (2010, 2 June). Japanese PM Yukio Hatoyama resigns amid Okinawa row. BBC News. Retrieved from www.bbc.com/news/10211314

Bulkeley, R. (2010). The political origins of the Antarctic Treaty. *Polar Record*, *46*(1), 9–11.

Burrett, T. (2014). Reaffirming Russia's remote control: exploring Kremlin influence on television coverage of Russian–Japanese relations and the southern Kuril Islands territorial dispute. *Demokratizatsiya*, *22*(3), 359–381.

Butcher, M. (2012) *'. . . When the long trick's over': Donald Kennedy in the Pacific*. Bendigo, VIC: Holland House.

Butler, J. (2002, 1 April). Guantanamo limbo: international law offers too little protection for prisoners of the new war. *Nation-New York*, *274*(12), 20–23.

Byers, M. (2010). *Who owns the Arctic? Understanding sovereignty disputes in the North*. Madeira Park, BC: Douglas & McIntyre.

Byers, M. (2013). *International law and the Arctic*. Cambridge: Cambridge University Press.

Byers, M. (2014). Creative thinking on sovereignty. *Policy Options*. Retrieved from http://policyoptions.irpp.org/issues/opening-eyes/byers/

Cai, Hong (2014, 16 December). Parliament win gives Abe a free hand. *China Daily*. Retrieved from http://usa.chinadaily.com.cn/opinion/2014-12/16/content_19094430.htm

Cairo Declaration (1943). Retrieved from www.ndl.go.jp/constitution/e/shiryo/01/002_46/002_46tx.html

Caldwell, D. (1996). Flashpoints in the Gulf: Abu Musa and the Tunb islands. *Middle East Policy 4*(3), 50–57.

Caldwell, L. K. (1998). *The national environmental policy act: An agenda for the future*. South Bend, IN: Indiana University Press.

Calvert, P. (1994). *The international politics of Latin America*. Manchester: Manchester University Press.

Cameron, A. (2012). Splendid isolation: 'philosopher's islands' and the reimagination of space. *Geoforum*, *43*(4), 741–749.

Campbell, D. (2006, 12 July). Vanuatu tops wellbeing and environment index. *Guardian*. Retrieved from www.theguardian.com/world/2006/jul/12/healthandwellbeing.lifeandhealth

Carney, M. (2016, 11 March). Is China's Xi Jinping more powerful than Mao? *ABC News*. Retrieved from www.abc.net.au/7.30/content/2015/s4423685.htm

Cassola, A. (1995). *The great siege of Malta (1565) and the Istanbul state archives*. Valletta: Said International.

Castagna, R. G., Thornton, L. L. and Tyrawski, J. M. (1999). *GIS and coastal boundary disputes: Where is Ellis Island?* ESRI Proceedings. Retrieved from http://proceedings.esri.com/library/userconf/proc99/proceed/papers/pap138/p138.htm

CBC News (2005a, 19 September). Canada, Denmark agree to resolve dispute over Arctic island. Retrieved from www.cbc.ca/news/world/canada-denmark-agree-to-resolve-dispute-over-arctic-island-1.551223

CBC News (2005b, 11 October). Disputing boundaries. Retrieved from www.cbc.ca/news2/background/boundary_disputes/

Chalfin, B. (2015). Governing offshore oil: mapping maritime political space in Ghana and the Western Gulf of Guinea. *South Atlantic Quarterly*, *114*(1), 101–118.

Chambers, K. and Chambers, A. (2000). *Unity of heart: Culture and change in a Polynesian atoll society*. Long Grove, IL: Waveland Press.

Charlton, M. (1989). *The little platoon: Diplomacy and the Falklands dispute*. Oxford: Basil Blackwell.

Charney, J. I. and Alexander, L. M. (eds) (1991). *International maritime boundaries*. The Hague: Martinus Nijhoff.

Chayes, A. (1974). *The Cuban missile crisis*. New York: Oxford University Press.

Chee, Yew Cheang (2013, 9 July). Transocean sets world record for deepwater drilling. *Rigzone*. Retrieved from www.rigzone.com/news/oil_gas/a/127610/Transocean_Sets_World_Record_for_Deepwater_Drilling

Chen, Boyu and Hwang, San Yih (2015). Senkaku/Diaoyu islands dispute and Taiwan's netizens' sentiments toward Japan and China. *East Asia*, *32*(4), 385–399.

Cheng, Tao (1973–1974). The Sino-Japanese dispute over the Tiao-yu-tai (Senkaku) islands and the law of territorial acquisition. *Virginia Journal of International Law*, *14*(2), 221–266.

Chen Weiss, J. (2014). *Powerful patriots: Nationalist protest in China's foreign relations*. Oxford: Oxford University Press.

Chermayeff, I., Wasserman, F. and Shapiro, M. J. (eds) (1991). *Ellis Island: An illustrated history of the immigrant experience*. New York: Macmillan.

China Military News (2010, 26 November). *Heixiazi island to become major joint China–Russian tourist resort*. Retrieved from www.china-defense-mashup.com/heixiazi-island-to-become-major-joint-china-russian-tourist-resort.html

Chinasmack.com (2010, 25 September). Chinese fishing boat captain released by Japanese, reactions. Retrieved from www.chinasmack.com/2010/stories/chinese-fishing-boat-captain-released-by-japanese-reactions.html

Chinese Society of International Law (2016, 10 June). The Tribunal's award in the 'South China Sea Arbitration' initiated by the Philippines is null and void. *China Daily*. Retrieved from www.chinadaily.com.cn/china/2016-06/10/content_25666477.htm

Choong, W. (2014). Events at a glance. *Adelphi Papers*, *54*(445), 9–12.

Christensen, T. J. (2002). The contemporary security dilemma: deterring a Taiwan conflict. *The Washington Quarterly*, *25*(4), 5–21.

Christensen, T. J. (2011). The advantages of an assertive China: responding to Beijing's abrasive diplomacy. *Foreign Affairs*, March/April. Retrieved from www.foreignaffairs.com/articles/east-asia/2011-02-21/advantages-assertive-china

Christensen, T. J. (2015a). *The China challenge: Shaping the choices of a rising power*. New York: W. W. Norton.

Christensen, T. J. (2015b). Obama and Asia: confronting the China challenge. *Foreign Affairs,* September/October, 28–36. Retrieved from www.foreignaffairs.com/articles/asia/obama-and-asia

Chun, Han Wong (2015, 8 February). From guns to cameras: Chinese tourism transforms Taiwan's Kinmen. *Wall Street Journal.* Retrieved from www.wsj.com/articles/from-guns-to-cameras-chinese-tourism-transforms-taiwans-kinmen-1423448694

Chung, C. (2004). *Domestic politics, international bargaining and China's territorial disputes*. New York: Routledge.

Churchill, R. R. (1985). Maritime delimitation in the Jan Mayen area. *Marine Policy*, *9*(1), 16–38.

CIA (2015). *The world factbook.* Washington, DC: Central Intelligence Agency. Retrieved from www.cia.gov/library/publications/the-world-factbook/

Clarke, H. D., Mishler, W. and Whiteley, P. (1990). Recapturing the Falklands: models of Conservative popularity, 1979–83. *British Journal of Political Science*, *20*(1), 63–81.

CLCS (2016). *Submissions, through the Secretary-General of the United Nations, to the Commission on the Limits of the Continental Shelf, pursuant to article 76, paragraph 8, of the United Nations Convention on the Law of the Sea of 10 December 1982*. New York: United Nations, Division for Ocean Affairs and the Law of the Sea. Retrieved from www.un.org/depts/los/clcs_new/commission_submissions.htm

Cleary, P. (2007). *Shakedown: Australia's grab for Timor oil* (Vol. 81). Crows Nest, NSW: Allen & Unwin.

CNN (2004, 7 August). Chinese riot after Japan win final. Retrieved from www.cnn.com/2004/SPORT/football/08/07/china.japan/

Cohen, J. A. (2013, 1 August). How dangerous are Sino-Japanese tensions? *China File*. Retrieved from www.chinafile.com/how-dangerous-are-sino-japanese-tensions

Conference on Antarctica (1959). *The Antarctic Treaty*. Retrieved from www.ats.aq/documents/ats/treaty_original.pdf

Connell, J. (2014). Alderney: gambling, Bitcoin and the art of unorthodoxy. *Island Studies Journal*, *9*(1), 69–78.

Convey, P., Hughes, K. A. and Tin, T. (2012). Continental governance and environmental management mechanisms under the Antarctic Treaty System: sufficient for the biodiversity challenges of this century? *Biodiversity*, *13*(3–4), 234–248.

Cooper, A. F. (2011). *Internet gambling offshore: Caribbean struggles over casino capitalism*. New York: Palgrave Macmillan.

Cordesman, A. H. (2007). *Iran, oil, and the Strait of Hormuz*. Washington, DC: Center for Strategic and International Studies. Retrieved from http://csis-prod.s3.amazonaws.com/s3fs-public/legacy_files/files/media/csis/pubs/070326_iranoil_hormuz.pdf

Costello, J. (1982). *The Pacific war: 1941–1945*. New York: HarperCollins.

Coutaz, G. (2015, May). China, confidence-building measures, and the South China Sea Peace Initiative. *South China Sea Think Tank*. Retrieved from http://scstt.org/issue-briefings/china-confidence-building-measures-and-the-south-china-sea-peace-initiative/

Coutinho, F. P. and Briosa e Gala, F. (2014). David and Goliath revisited: a tale about the Timor-Leste/Australia Timor Sea agreements. *Texas Journal of Oil, Gas & Energy Law*, *10*(2), 429–462. Retrieved from http://tjogel.org/journalarchive/Issue10/10.2.David Goliath_Francisco.pdf

CPSL (2010). *The role of spatial planning and sediment in coastal risk management*, CPSL Third Report, Wadden Sea Ecosystem No. 28. Common Wadden Sea Secretariat,

Trilateral Working Group on Coastal Protection and Sea Level Rise (CPSL). Wilhelmshaven, Germany.

Critchlow, A. (2015, 28 May). New Falklands oil discovery could stir trouble with Argentina. *Telegraph*. Retrieved from www.telegraph.co.uk/finance/newsbysector/energy/oilandgas/11635627/New-Falklands-oil-discovery-could-stir-trouble-with-Argentina.html

Cronin, P. M., Dutton, P. A., Fravel, M. T., Holmes, J. R., Kaplan, R. D., Rogers, W. and Storey, I. (2012). *Cooperation from strength: The United States, China and the South China Sea*. Washington, DC: Center for a New American Security.

CWSS (1998). *Stade Declaration: Trilateral Wadden Sea Plan*. Wilhelmshaven: Common Wadden Sea Secretariat.

CWSS (2008). *The Wadden Sea Waddenzee – Wattenmeer – Nomination of the Dutch–German Wadden Sea as World Heritage Site*. Wilhelmshaven: Common Wadden Sea Secretariat (CWSS).

CWSS (2010). *Wadden Sea Plan 2010*. Wilhelmshaven: Common Wadden Sea Secretariat (CWSS).

CWSS (2012). *The Wadden Sea, Germany and Netherlands (N1314) – Extension Denmark and Germany*. Wilhelmshaven: Common Wadden Sea Secretariat (CWSS).

CWSS (2014). *Sustainable tourism in the Wadden Sea World Heritage destination*. Wilhelmshaven: Common Wadden Sea Secretariat (CWSS).

CWSS (2016). Wadden Sea World Heritage Site Publications. Common Wadden Sea Secretariat. Retrieved from www.waddensea-secretariat.org/management/whs/whs.html

Dai Tan, J. D. (2006). The Diaoyu/Senkaku dispute: bridging the cold divide. *Santa Clara Journal of International Law*, *5*(1), 134–162.

Daly, M. W. (2003). *Imperial Sudan: The Anglo-Egyptian condominium, 1934–1956*. Cambridge: Cambridge University Press.

Dator, J. (2009). Alternative futures at the Manoa School. *Journal of Futures Studies*, *14*(2), 1–18.

Davis, S. P. (2015). *The empire's edge: Militarisation, resistance and transcending hegemony in the Pacific*. Atlanta, GA: University of Georgia Press.

Dean, M. and Nagashima, M. (2007). Sharing the burden: the role of government and NGOs in protecting and providing for asylum seekers and refugees in Japan. *Journal of Refugee Studies*, *20*(3), 481–508.

De Beaufort, L. F. (1954). *Veranderingen in de flora en fauna der Zuiderzee (thans IJsselmeer) na de afsluiting in 1932*. Den Helder: De Boer (in Dutch).

De Jonge, V. N. (2009). From a defensive to an integrated approach. In S. Reinhard and H. Folmer (eds), *Water policy in the Netherlands: Integrated management in a densely populated area* (pp. 17–46). Washington, DC: Resources for the Future Press.

De Jonge, V. N., Essink, K. and Boddeke, R. (1993). The Dutch Wadden Sea: a changed ecosystem. *Hydrobiologica*, *265*(1), 45–71.

De Jonge, V. N., Pinto, R. and Turner, R. K. (2012). Integrating ecological, economic and social aspects to generate useful management information under the EU Directives' ecosystem approach. *Ocean and Coastal Management*, *68*(1), 169–188.

De Meyer, D. (2011). Growing an island: Okinotori. *San Rocco*, *1*(1), 23–28. Retrieved from https://www.researchgate.net/publication/266597810_GrowinG_an_island_okinotori

DeLamater, E. (2011). *White college boy steel bands in 1950s Trinidad: How middle-class teenagers helped the steelpan gain national acceptance*. Doctoral dissertation. Phoenix, AZ: Arizona State University.

DeLoughrey, E. M. (2007). *Routes and roots: Navigating Caribbean and Pacific island literatures*. Honolulu, HI: University of Hawai'i Press.

DeLoughrey, E. M. (2013). The myth of isolates: ecosystem ecologies in the nuclear Pacific. *Cultural Geographies*, *20*(2), 167–184.

Denmark–Canada Treaty. (1973, 17 December). Agreement between the Government of the Kingdom of Denmark and the Government of Canada relating to the Delimitation of the Continental Shelf between Greenland and Canada. New York: United Nations. Retrieved from www.un.org/Depts/los/LEGISLATIONANDTREATIES/PDFFILES/TREATIES/DNK-CAN1973CS.PDF

Di Castri, F. and Balaji, V. (eds) (2002). *Tourism, biodiversity and information*. Leiden: Backhuys.

Dillon, G. M. (1989). *The Falklands, politics and war*. New York: St. Martin's Press.

Ding, Gang (2013, 12 June). Japan nervous about future of Asia-Pacific. *Global Times (China)*. Retrieved from www.globaltimes.cn/content/788415.shtml

Dittmer, L. (2015). China's new leadership and global role. In E. Berman and M. S. Haque (eds), *Asian leadership in policy and governance* (pp. 21–43). New York: Emerald Group Publishing.

Dodds, K. J. (2011). Sovereignty watch: claimant states, resources, and territory in contemporary Antarctica. *Polar Record*, *47*(3), 231–243.

Dodds, K. J. (2010). Governing Antarctica: contemporary challenges and the enduring legacy of the 1959 Antarctic Treaty. *Global Policy*, *1*(1), 108–115.

Dodds. K. J. and Royle S. A. (2003). The historical geography of islands. Introduction: rethinking islands. *Journal of Historical Geography*, *29*(4), 487–498.

Dolighan Cartoons (2005). Hans island. Retrieved from www.dolighan.com/image-view.php?image=jul2605c.jpg&PHPSESSID=2157d7d58ab67f9a3d171544d6d616a6

Donaghy, A. (2014). *The British Government and the Falkland Islands, 1974–79*. London: Palgrave Macmillan.

Dorow, S. K. (2006). *Transnational adoption: A cultural economy of race, gender, and kinship*. New York: NYU Press.

Downs, E. S. and P. C. Saunders (2000). Legitimacy and the limits of nationalism: China and the Diaoyu Islands. In M. E. Brown (ed.), *The rise of China* (pp. 41–73). Cambridge, MA: MIT Press.

Dreyer, J. T. (2015). The rise of China and the geopolitics of East Asia. *Orbis*, *59*(4), 518–529..

Drifte, R. (2009). Territorial conflicts in the East China Sea: from missed opportunities to negotiation stalemate. *Asia Pacific Journal: Japan Focus*, *7*(22), No. 3. Retrieved from http://apjjf.org/-Reinhard-Drifte/3156/article.pdf

Drifte, R. (2014, 28 July). The Japan–China confrontation over the Senkaku/Diaoyu islands: between 'shelving' and 'dispute escalation'. *The Asia-Pacific Journal*, *12*, Issue 30(3). Retrieved from www.japanfocus.org/-Reinhard-Drifte/4154

Drower, G. M. (2002). *Heligoland: The true story of German Bight and the island that Britain betrayed*. Stroud: Sutton Publishers.

Dudden, A. (2015, 16 January). The shape of Japan to come. *New York Times*. Retrieved from www.nytimes.com/2015/01/17/opinion/the-shape-of-japan-to-come.html

Dutton, P. A. (2014, 14 January). *China's maritime disputes in the East and South China seas*. Testimony before a hearing of the House Foreign Affairs Committee. Retrieved from www.usnwc.edu/getattachment/9edbcea9-8425-4b96-aa14-aac1f81532c2/China-s-Maritime-Disputes-in-the-East-and-South-Ch.aspx

Dyer, G. (2013, 27 January). Senkaku/Diaoyu: another Falklands? Personal blog. Retrieved from www.gwynnedyer.com/tag/argentina/

ECSPI (2012, 15 August). *East China Sea Peace Initiative*. Taipei: Ministry of Foreign Affairs, Republic of China. Retrieved from www.mofa.gov.tw/en/theme.aspx?s=780E70E6D142B833&sms=BCDE19B435833080

Eddy, P. and Linklater, M. (1982). *The Falklands War: The full story by the Sunday Insight team*. London: Sphere Books.

Edenhofer, O., Flachsland, C., Jakob, M. and Lessmann, K. (2013). The atmosphere as a global commons: challenges for international cooperation and governance. Discussion Paper 2013-58. Cambridge MA: Harvard Project on Climate Agreements. Retrieved from http://belfercenter.hks.harvard.edu/files/hpcadp58_edenhofer-flachsland-jakob-lessmann.pdf

Edgar, G. J., Russ, G. R. and Babcock, R. C. (2007). Marine protected areas. In S. D. Connell and B. M. Gillanders (eds), *Marine ecology* (pp. 533–555). South Melbourne, VIC: Oxford University Press.

Edmond, R. and Smith, V. (eds) (2003). *Islands in history and representation*. London: Routledge.

Eilperin, J. (2014, 27 April). U.S., Philippines reach 10-year defence agreement amid rising tensions. *Washington Post*. Retrieved from www.washingtonpost.com/world/us-philippines-to-sign-10-year-defense-agreement-amid-rising-tensions/2014/04/27/a04436c0-cddf-11e3-a75e-463587891b57_story.html

Elango, L. Z. (1987). *The Anglo-French condominium in Cameroon, 1914–1916: History of a misunderstanding*. Yaounde: Navi-Group Publications.

Elliott, L. (2004). *The global politics of the environment*. New York: Palgrave Macmillan.

Emery, K. O., Hayashi, Y., Hilde, T. W., Kobayashi, K., Koo, J. H., Meng, C. Y., Niino, H. . . . and Yang, S. J. (1969). Geological structure and some water characteristics of the East China Sea and the Yellow Sea. *Coordinating Committee for Geoscience Programmes in East and Southeast Asia (CCOP) Technical Bulletin*, Vol. 2, 3–43.

Estevez-Abe, M. (2014). Feeling triumphalist in Tokyo: the real reasons nationalism is back in Japan. *Foreign Affairs*, *93*, May/June. Retrieved from www.foreignaffairs.com/reviews/review-essay/feeling-triumphalist-tokyo

Evans, A. (2015, 17 May). Okinawa demonstrations and Osprey deployment to Yokota US military base in Japan increase protest risk in Tokyo. *IHS Jane's 360*. Retrieved from www.janes.com/article/51489/okinawa-demonstrations-and-osprey-deployment-to-yokota-us-military-base-in-japan-increase-protest-risk-in-tokyo

Evans, H. J. (2014a, 13 March). 'What if' is a waste of time. *Guardian*. Retrieved from www.theguardian.com/books/2014/mar/13/counterfactual-history-what-if-waste-of-time

Evans, H. J. (2014b, 28 June). Debate: is 2014, like 1914, a prelude to world war? *The Globe and Mail*, F4.

Evans, M. D. and Shaw, M. N. (1993). Case concerning the land, island and maritime frontier dispute (El Salvador/Honduras: Nicaragua intervening), Judgment of 11 September 1992. *International and Comparative Law Quarterly*, *42*(4), 929–937.

Everard, J. and Holt, J. C. (2004). *Jersey 1204: the forging of an island community*. London: Thames & Hudson.

Everingham, S. (2015, 3 June). East Timor revives attempt to have Greater Sunrise treaty declared invalid amid spying allegations. *Australia Broadcasting Corporation*. Retrieved from www.abc.net.au/news/2015-06-03/row-over-east-timor-oil-and-gas-treaty-reignites/6519722

Fackler, M. (2015a, 22 February). Security guards detain three protesters at gate of marine base on Okinawa. *New York Times*. Retrieved from www.nytimes.com/2015/02/23/world/asia/marine-base-okinawa.html

Fackler, M. (2015b, 1 April). Japan ends decades-long ban on export of weapons. *New York Times*. Retrieved from www.nytimes.com/2014/04/02/world/asia/japan-ends-half-century-ban-on-weapons-exports.html

Fackler, M. and Johnson, I. (2010, 24 September). Japan retreats with release of Chinese boat captain. *New York Times*. Retrieved from www.nytimes.com/2010/09/25/world/asia/25chinajapan.html?_r=0

Fairbank, J. K. (1953). *Trade and diplomacy on the China coast: The opening of the treaty ports, 1842–1854* (Vol. I). Cambridge MA: Harvard University Press.

Farbotko, C. and Lazrus, H. (2012). The first climate refugees? Contesting global narratives of climate change in Tuvalu. *Global Environmental Change*, *22*(2), 382–390.

Fellman, K., Kinnunen, J., Lindström, B. and Palmer, R. (2015). Pathways to successful entrepreneurship in small island economies: the case of Åland. In G. Baldacchino (ed.), *Entrepreneurship in small island states and territories* (pp. 65–79). New York: Routledge.

Ferguson, N. (2006, 26 June). The triumph of the east. *New Statesman*. Retrieved from www.newstatesman.com/node/153510

Ficenec, J. (2014, 4 September). Argentina loses squid war with Falklands. *Telegraph*. Retrieved from /www.telegraph.co.uk/finance/11074615/Argentina-loses-squid-war-with-Falklands.html

Fischer, L. (2011). Küste – von der realität eines mentalen Konzepts. In L. Fischer and K. Reise (eds), Küstenmentalität und Klimawandel. *Oekom* (pp. 31–53). Munich (in German).

Flamm, P. (2014, 26 August). Dokdo and South Korean national identity. *The Diplomat*. Retrieved from http://thediplomat.com/2014/08/dokdo-and-south-korean-national-identity/

Fleury, C. (2006). Discontinuités et systèmes spatiaux: la combinaison île/frontière à travers les exemples de Jersey, de Saint-Pierre-et-Miquelon et de Trinidad. PhD thesis, Université de Caen Basse-Normandie (in French).

Fleury, C. (2011). Jersey and Guernsey: two distinct approaches to cross border fishery management. *Shima: The International Journal of Research into Island Cultures*, *5*(1), 24–43.

Foley, S. (1999). The UAE: political issues and security dilemmas. *Middle East Review of International Affairs*, 3(1). Retrieved from www.rubincenter.org/meria/1999/03/foley.pdf

Foot, R. (2014). 'Doing some things' in the Xi Jinping era: the United Nations as China's venue of choice. *International Affairs*, *90*(5), 1085–1100.

Forster, R. A. S. (1980). Vanuatu: the end of an episode of schizophrenic colonialism. *The Round Table: Commonwealth Journal of International Affairs*, *70*(280), 367–373.

Foster, J. B. (2006). *Naked imperialism: The U.S. pursuit of global dominance*. New York: Monthly Review Press.

Foster, M. (2012, 28 August). Japan leader sends letter to Beijing amid tension. Associated Press. Retrieved from http://bigstory.ap.org/article/car-carrying-japan-ambassador-attacked-beijing

Franck, T. M. (1995). *Fairness in international law and institutions*. Oxford: Clarendon Press.

François, A. (2012). The dynamics of linguistic diversity: egalitarian multilingualism and power imbalance among northern Vanuatu languages. *International Journal of the Sociology of Language*, *214*, 85–110.

Frankel, J. (2010). Mauritius: African success story. *Working Paper Series RWP10-036*. Cambridge, MA: Harvard University, Kennedy School of Government. Retrieved from file:///C:/Users/default.SID24596/Downloads/RWP10-036_Frankel.pdf

Fravel, M. T. (2012, 28 September). Something to talk about in the East China Sea. *The Diplomat: China Power*. Retrieved from http://thediplomat.com/2012/09/something-to-talk-about-in-the-east-china-sea/

Free Hans Island (2015). The official website of the Free Hans Island Movement. Retrieved from http://freehansisland.com/

Freedman, L. and Stonehouse-Gamba, V. (1982). *Signals of war: The Falklands crisis*. London: Faber and Faber.

Friedman, N. (2015, 6 August). Oil prices near six-year lows on oversupply fears. *Wall Street Journal*. Retrieved from www.wsj.com/articles/crude-oil-futures-lose-momentum-1438830721

Fuentes, G. (2015, 5 September). Navy signs off on plan to move 5,000 Marines to Guam. *Marine Corps Times*. Retrieved from www.marinecorpstimes.com/story/military/2015/09/05/navy-signs-off-plan-move-5000-marines-guam/71657614/

Gady, F.-S. (2015a, 28 March). Japan's new helicopter carrier: bad news for Chinese subs? *The Diplomat*. Retrieved from http://thediplomat.com/2015/03/japans-new-helicopter-carrier-bad-news-for-chinese-subs/

Gady, F.-S. (2015b, 6 May). Meet Japan's newest aircraft for defending the Senkaku islands. *The Diplomat*. Retrieved from http://thediplomat.com/2015/05/meet-japans-newest-aircraft-for-defending-the-senkaku-islands/

Gai, Qi (2015). Image reconstruction and the reflection of values in the formation of national traumatic memories: a review of recent anti-Japanese war films and teleplays in China. *Frontiers of Literary Studies in China*, *9*(2), 306–317.

Gammeltoft-Hansen, T. (2014). Extraterritorial migration control and the reach of human rights. In V. Chetail and C. Bauloz (eds), *Research handbook on international law and migration* (pp. 113–132). Cheltenham: Edward Elgar.

Gelber, H. (2004). *Opium, soldiers and evangelicals: England's 1840–42 war with China and its aftermath*. New York: Palgrave Macmillan.

Gelernter, J. (2016, 16 July). Japan reverts to fascism. *National Review*. Retrieved from www.nationalreview.com/article/437950/japans-new-fascism

Genba, K. (2012, 20 November). Japan–China relations at a crossroads. *New York Times*. Retrieved from www.nytimes.com/2012/11/21/opinion/koichiro-genba-japan-china-relations-at-a-crossroads.html?_r=0

Gensini, G., Yacoub, M. and Conti, A. (2004). The concept of quarantine in history: from plague to SARS. *Journal of Infection*, *49*(4): 257–261.

Ghai, Y. P. (2000). *Autonomy and ethnicity: Negotiating competing claims in multi-ethnic states*. Cambridge: Cambridge University Press.

Giacomo, C. (2014, 20 April). Focusing America's attention on Asia. *New York Times*. Retrieved from www.nytimes.com/2014/04/21/opinion/focusing-americas-attention-on-asia.html?_r=0

Giebels, D., van Buuren, A. and Edelenbos, J. (2013). Ecosystem-based management in the Wadden Sea: principles for the governance of knowledge. *Journal of Sea Research*, *82*, 176–187.

Gilbert, M. (2004). *The Second World War: A complete history*. Chapel Hill, NC: University of North Carolina Press.

Gillard, D. R. (1960). Salisbury's African policy and the Heligoland offer of 1890. *The English Historical Review*, *75*(297), 631–653.

Glasscock, J. (1985). *The making of an island: Sint Maarten/Saint Martin*. Wellesley, MA: The Windsor Press.

Goldsmith, W. W. (1979). Operation Bootstrap, industrial autonomy and a parallel economy for Puerto Rico. *International Regional Science Review*, *4*(1), 1–22.

Goldstein, L. J. (2001). Who started shooting and why it matters. *The China Quarterly*, *168*, 985–997.

Good, T. P., Tuttle, V., Jannot, J., Shama, R., Riley, N. and McVeigh, J. (2015). *Observed and estimated by-catch of short-tailed albatross in U.S. West Coast groundfish fisheries, 2010–2013*. Seattle, WA: Fisheries Observation Science Program, National Marine Fisheries Service. Retrieved from www.pcouncil.org/wp-content/uploads/2015/06/D4_Sup_Att5_Bycatch_Rpt_short-tailed-albatross_E-ONLY_JUN2015BB.pdf

Greenhouse, L. (1998, 27 May). The Ellis Island verdict: the ruling; high court gives New Jersey most of Ellis Island. Retrieved from www.nytimes.com/1998/05/27/nyregion/ellis-island-verdict-ruling-high-court-gives-new-jersey-most-ellis-island.html

Gries, P. H. (2004). *China's new nationalism: Pride, politics and diplomacy*. Berkeley, CA: University of California Press.

Grob, J. (2007). Antarctica's frozen territorial claims: a meltdown proposal. *Boston College International & Comparative Law Review*, *30*(2), 461–484.

Groh, G. (2014). Gustavo Groh: El agua apagó el fuego. Retrieved from www.gustavogroh.net/portfolio/el-agua-apago-el-fuego/

Grubb, J. W. (1999). *British and French administration of the New Hebrides, 1904–1910: condominium or pandemonium?* Unpublished dissertation. Otago: University of Otago, Department of History.

Grydehøj, A. (2014). Informal diplomacy in Norway's Svalbard policy: the intersection of local community development and Arctic international relations. *Global Change, Peace & Security*, *26*(1), 41–54.

Grydehøj, A., Grydehøj, A. and Ackrén, M. (2012). The globalisation of the Arctic: negotiating sovereignty and building communities in Svalbard, Norway. *Island Studies Journal*, *7*(1), 99–118.

Guadeloupe, F. (2009). *Chanting down the New Jerusalem: Calypso, Christianity and capitalism in the Caribbean*. Berkeley, CA: University of California Press.

Guardian (2016, 9 June). Japan protests after Chinese warship sails near disputed islands. Retrieved from www.theguardian.com/world/2016/jun/09/japan-protests-after-chinese-warship-sails-near-disputed-islands

Guibernau, M. (2013). *Nationalisms: The nation-state and nationalism in the twentieth century*. New York: John Wiley & Sons.

Gullett, W. (2016). Reconciliation in the Timor Sea. *The Korean Journal of International and Comparative Law*, *4*(1), 99–112.

Guo, Rongxing (2010). *Territorial disputes and seabed petroleum exploitation: Some options for the East China Sea*. Washington, DC: The Brookings Institution: Center for Northeast Asian Policy Studies. Retrieved from www.brookings.edu/~/media/research/files/papers/2010/9/east%20china%20sea%20guo/09_east_china_sea_guo.pdf

Guo, Rongxing (2015). Alternative strategies for dispute resolution. In *Cross-Border management* (pp. 179–203). Berlin: Springer.

Gupta, P. (2015). Futures, fakes and discourses of the gigantic and miniature in 'The World' islands' development, Dubai. *Island Studies Journal, 10*(2), 181–196.

Haas, P. (2011). Status and sovereignty of the Liancourt rocks: the dispute between Japan and Korea, *Gonzaga Journal of International Law, 2*, 2–28.

Haase, D., Lamers, M. and Amelung, B. (2009). Heading into uncharted territory? Exploring the institutional robustness of self-regulation in the Antarctic tourism sector. *Journal of Sustainable Tourism, 17*(4), 411–430.

Halpin, D. P. (2015, 9 March). U.S. should be appalled by Japan's historical revisionism. *The National Interest*. Retrieved from http://nationalinterest.org/feature/us-should-be-appalled-by-japans-historical-revisionism-12381

Hara, K. (2012, 22 April). The San Francisco Peace Treaty and frontier problems in the regional order in East Asia: a 60-year perspective. *The Asia Pacific Journal | Japan Focus, 10*(17)(1), 1–17. Retrieved from http://apjjf.org/-Kimie-Hara/3739/article.pdf

Hara, K. (2015). Okinawa, Taiwan and the Senkaku/Diaoyu islands in the United States–Japan–China relations. In T. F. Liao, K. Hara and K. Wiegand (eds), *The China–Japan border dispute: Islands of contention in multidisciplinary perspective* (pp. 37–57). Farnham: Ashgate.

Hardin, G. (1968). The tragedy of the commons. *Science, 162*(3859), 1243–1248.

Harding, R. (2016, 10 July). Shinzo Abe wins sweeping victory in Japan elections. *Financial Times*. Retrieved from https://next.ft.com/content/d7982268-4668-11e6-8d68-72e9211e86ab

Hardy, J. and O'Connor, S. (2014, 20 November). China building airstrip-capable island on Fiery Cross Reef. *IHS Jane's Defence Weekly*. Retrieved from www.janes.com/article/46083/china-building-airstrip-capable-island-on-fiery-cross-reef

Harlan, C. (2011, 13 December). Japan to pick Lockheed's F-35 as new stealth fighter. *The Washington Post*. Retrieved from www.washingtonpost.com/world/japan-to-pick-lockheeds-f-35-as-new-stealth-fighter/2011/12/13/gIQAbuYUrO_story.html

Harnisch, S. (2016). China's historical self and its international role. In S. Harnisch, S. Bersick and J.-C. Gottwald (eds), *China's international roles: Challenging or supporting international order?* (pp. 38–58). New York: Routledge.

Harris, J. J. (2013). The Pacific war, continued: denationalising international law in the Senkaku/Diaoyu island dispute. *Georgia Journal of International and Comparative Law, 42*, 587–613.

Harris, P. (2015). America's other Guantánamo: British foreign policy and the US base on Diego Garcia. *The Political Quarterly, 86*(4), 507–514.

Harsson, B. G. and Preiss, G. (2014). Norwegian baselines, maritime boundaries and the UN Convention on the Law of the Sea. *Arctic Review on Law and Politics, 3*(1), 108–129.

Hartog, J. (1981). *History of Sint Maarten and Saint Martin*. Philipsburg, Sint Maarten: The Sint Maarten Jaycees.

Hatton, C. (2014, 23 April). Would China fight for disputed islands? BBC News. Retrieved from www.bbc.com/news/blogs-china-blog-27122084

Heflin, W. B. (2000). Diaoyu/Senkaku islands dispute: Japan and China, oceans apart. *Asian-Pacific Law and Policy Journal, 18*(1) 1–22.

Hellyer, R. I. (2005). The missing pirate and the pervasive smuggler: regional agency in coastal defence, trade, and foreign relations in nineteenth-century Japan. *International History Review, 27*(1), 1–24.

Hemmings, A. D., Chaturvedi, S., Leane, E., Liggett, D. and Francisco, J. F. (2014). The role of nationalism in the contemporary Antarctic. In *Proceedings of the 7th Polar*

Law symposium: The Polar regions in the Asian century. Hobart. Retrieved from www.researchgate.net/publication/270396980_The_Role_of_Nationalism_in_the_Contemporary_Antarctic

Hepburn, E. (2014). Forging autonomy in a unitary state: the Åland Islands in Finland. *Comparative European Politics*, *12*(4), 468–487.

Hernández, J. C. (2015). Warily eyeing China, Philippines may invite U.S. back to Subic Bay. *New York Times*. Retrieved from www.nytimes.com/2015/09/20/world/warily-eyeing-china-philippines-may-invite-us-back-to-subic-bay.html?_r=0

Hillebrink, S. (2013). Saint-Martin/Sint Maarten. In G. Baldacchino (ed.), *The political economy of divided islands: Unified geographies, multiple polities* (pp. 176–194). London: Palgrave Macmillan.

Hillebrink, S. (2008). *The right to self-determination and post-colonial governance: The case of the Netherlands Antilles and Aruba*. Cambridge: Cambridge University Press.

Hillebrink, S. and Bense, M. (2008). Zijn de Antillen te koop? In P. van Sasse van Ysselt (ed.), *Handhaving van de openbare orde: is voorkomen beter dan genezen? Bundel ter gelegenheid van het afscheid van prof. mr. A.E. Schilder* (pp. 105–115). The Hague: Directie Constitutionele Zaken en Wetgeving van het Ministerie van Binnenlandse Zaken en Koninkrijksrelaties (in Dutch).

Hills, T. L. (1996). Island matters, islands matter: international institutional responses to the analysis of problems and their solutions. *Geographische Zeitschrift*, *84*(2), 67–73.

Ho, J. (2010). The implications of Arctic sea ice decline on shipping. *Marine Policy*, *34*(3), 713–715.

Hobsbawm, E. (1983). Falklands fallout. *Marxism Today*, *9*, 13–19. Retrieved from www.amielandmelburn.org.uk/collections/mt/pdf/83_01_13.pdf

Hoeneveld, A. (2002). Het Verdrag van Concordia van 1648. In O. Balak, R. G. Mazel, A. E. Schilder, P. H. J. Smeets and M. J. Willemse (eds), *Wetten, woorden, wensen. Opstellen over constitutie, wetgeving en beleid* (pp. 89–97). The Hague: Directie Constitutionele Zaken en Wetgeving van het Ministerie van Binnenlandse Zaken en Koninkrijksrelaties (in Dutch).

Hoffmann, M. J. (2012). *Ozone depletion and climate change: Constructing a global response*. Ithaca, NY: SUNY Press.

Holmes, J. (2013, 15 April). Three reasons to applaud Taiwan-Japan fishing accord. *The Diplomat*. Retrieved from http://thediplomat.com/2013/04/3-reasons-to-applaud-taiwan-japan-fishing-accord/

Holmes, J. (2015, 23 September). When China rules the sea. *Foreign Policy*. Retrieved from http://foreignpolicy.com/2015/09/23/when-china-rules-the-sea-navy-xi-jinping-visit/?utm_content=buffer83813&utm_medium=social&utm_source=facebook.com&utm_campaign=buffer

Hook, G. D. (2014). Japan's risky frontiers: territorial sovereignty and governance of the Senkaku islands. *Japanese Studies*, *34*(1), 1–23.

Hornung, J. W. (2015, 4 May). U.S.–Japan: a Pacific alliance transformed. *The Diplomat*. Retrieved from http://thediplomat.com/2015/05/u-s-japan-a-pacific-alliance-transformed/

Hoskins, G. (2012). On arrival: memory and temporality at Ellis Island, New York. *Environment and Planning D*, *30*(6), 1011–1027.

Howe, J. (2013). *A review of Tuvalu's gross national income statistics*. Funafuti: Ministry of Finance and Economic Development, Department of Planning and Budget & Central Statistics Division, Tuvalu Government. July.

Howell, D. L. (1998). Territoriality and collective identity in Tokugawa Japan. *Daedalus*, *127*(3), 105–132.

Hsü, I. C. Y. (2000). *The rise of modern China*. New York: Oxford University Press.

Hsu, K. (2014, 14 January). *Air defense identification zone intended to provide China greater flexibility to enforce East China Sea claims*. U.S. –China Economic and Security Review Commission staff report. Retrieved from www.uscc.gov/sites/default/files/Research/China%20ADIZ%20Staff%20Report.pdf

Hu, Jintao (2012, 17 November). Full text of Hu Jintao's report at 18th Party Congress, section viii (English version). Retrieved from http://news.xinhuanet.com/english/special/18cpcnc/2012-11/17/c_131981259_9.htm

Hughes, C. W. (2015). *Japan's foreign and security policy under the 'Abe Doctrine': New dynamism or new dead end?* New York: Palgrave Macmillan.

Humphreys, A. (2012, 11 April). New proposal would see Hans Island split equally between Canada and Denmark. *National Post*. Retrieved from http://news.nationalpost.com/news/canada/new-proposal-would-see-hans-island-split-equally-between-canada-and-denmark

Humphries, C. and Ferguson, A. (2016, 5 July). How Brexit has suddenly made the prospect of a united Ireland thinkable. *Independent*. Retrieved from www.independent.co.uk/news/uk/politics/brexit-what-it-means-for-northern-ireland-united-eu-referendum-result-irish-passport-a7117166.html

Hyer, E. (1996). The Sino-Russian boundary settlement. *IBRU Boundary and Security Bulletin, 4*(2), 90–94. Retrieved from www.dur.ac.uk/resources/ibru/publications/full/bsb4-2_hyer.pdf

Hyer, E. (2015). *The pragmatic dragon: China's grand strategy and boundary settlements*. Vancouver, BC: UBC Press.

ICJ (1953). *Minquiers and Écréhous Case*. Judgment of 17 November 1953. The Hague: International Court of Justice. Retrieved from www.icj-cij.org/docket/files/17/2023.pdf

ICJ (1992). *Case concerning land, island and maritime frontier dispute (El Salvador/Honduras: Nicaragua Intervening)*. Judgment of 11 September 1992. The Hague: International Court of Justice. Retrieved from www.icj-cij.org/docket/index.php?sum=390&code=sh&p1=3&p2=3&case=75&k=0e&p3=5

ICJ (1999a). *Sovereignty over Kasikili/Sedudu island (Botswana/Namibia)*. The Hague: International Court of Justice. Retrieved from www.icj-cij.org/docket/index.php?sum=505&p1=3&p2=3&case=98&p3=5

ICJ (1999b, 13 December). ICJ finds that Kasikili/Sedudu island forms part of territory of Botswana. *United Nations Press Release ICJ/594*. Retrieved from www.un.org/news/Press/docs/1999/19991213.icj594.doc.html

ICJ (2002). *Sovereignty over Pulau Ligitan and Pulau Sipadan (Indonesia/Malaysia)*. The Hague: International Court of Justice. Retrieved from www.icj-cij.org/docket/index.php?p1=3&p2=3&k=df&case=102&code=inma&p3=5

ICJ (2008). *Sovereignty over Pedra Branca/Pulau Batu Puteh, Middle Rocks and South Ledge (Malaysia/Singapore)*. The Hague: International Court of Justice. Retrieved from www.icj-cij.org/docket/index.php?p1=3&p2=3&k=2b&case=130&code=masi&p3=5

IIATO (2015). *Tourism statistics*. International Association of Antarctica Tour Operators. Retrieved from http://iaato.org/documents/10157/1017626/Tourists+by+Nationality+-+Total.pdf/9a175577-5e15-4ee8-97d4-5a4327c4ea25

Ikehara, S. and Shimojana, M. (1971). Senkaku Retto no rikusei shokubutsu [Land animals of the Senkaku Islands]. In *Senkaku Retto gakujutsu chosa hokoku* [The Senkaku Islands Academic Survey Report]. Okinawa: University of the Ryukyus, July (in Japanese).

ILSA (2012). Judicial decisions involving questions of international law: France–Mexico. Arbitral award on the subject of the difference relative to the sovereignty over Clipperton

Island. International Law Students Association. Retrieved from www.ilsa.org/jessup/jessup10/basicmats/clipperton.pdf

International Herald Tribune (2008, 8 July). 900-year-old temple on disputed Thai–Cambodia border named World Heritage Site. Retrieved from https://web.archive.org/web/20090210122753/http://www.iht.com/articles/ap/2008/07/08/america/NA-Canada-Thailand-Cambodia-Temple.php

Ishiyama, E. (2015, 26 January). Ryukyu pro-independence group quietly gathering momentum. *Japan Times*. Retrieved from www.japantimes.co.jp/news/2015/01/26/national/ryukyu-pro-independence-group-quietly-gathering-momentum/#.VaV7h_kYM90

IUCN Red List (2016). Mogera uchidai (Ryukyu Mole, Senkaku Mole). Retrieved from www.iucnredlist.org/details/14563/0

Iwashita, A. (2004). *A 4,000 kilometre journey along the Sino-Russian border*. Sapporo: Slavic Research Centre.

Iwashita, A. (2005a). An inquiry for new thinking on the border dispute: backgrounds of 'historic success' for the Sino-Russian negotiations. In *Siberia and the Russian Far East in the 21st Century: Partners in the community of Asia, Vol. 1: Crossroads in Northeast Asia* (pp. 95–114). Sapporo: Slavic Research Centre.

Iwashita, A. (2005b). *Hoppo ryodo mondai* [Northern Territories Issue]. Tokyo: Chukoshinsho (in Japanese).

Iwashita, A. (2013). Bolshoi Ussuriiskii/Heixiazi. In G. Baldacchino (ed.), *The political economy of divided islands: Unified geographies, multiple polities* (pp. 212–226). New York: Palgrave Macmillan.

Jabour, J. (2010). Biological prospecting: the ethics of exclusive reward from Antarctic activity. *Ethics in Science and Environmental Politics*, *10*(1), 19–29. Retrieved from www.int-res.com/articles/esep2010/10/e010p019.pdf

Jackson, M. and della Dora, V. (2011). Spectacular enclosures of hope: artificial islands in the Gulf and the present. In R. Shields, O. Park and T. Davidson (eds), *Ecologies of affect: Placing nostalgia, desire and hope* (pp. 293–316). Waterloo, ON: Wilfrid Laurier University Press.

Jacobs, F. (2012, 23 January). Borderlines: the world's most exclusive condominium. *New York Times*. Retrieved from http://opinionator.blogs.nytimes.com/2012/01/23/the-worlds-most-exclusive-condominium/?_r=0#ftn14

Jacobsson, M. (1998). *Building the international legal framework for Antarctica*. Stockholm: Swedish Ministry of Foreign Affairs. Retrieved from www.atsummit50.org/media/book-5.pdf

James, H. (2014, 28 June). Is 2014, like 1914, a prelude to world war? *The Globe and Mail*, F4.

Japan Reference (2012, 25 May). The tragic fate of the Japanese albatross. Retrieved from www.jref.com/culture-society/japanese-albatross/

Japan Statistical Yearbook (2015). Islands, length of coastline and area of national land. Retrieved from www.stat.go.jp/data/nenkan/zuhyou/y0101000.xls

Japan Times (2014a, 30 March). Five dead after pier overturns off Japan's southernmost island. Retrieved from www.japantimes.co.jp/news/2014/03/30/national/five-dead-after-pier-overturns-off-japans-southernmost-island/#.VapQN_kYM90

Japan Times (2014b, 19 April). U.S. military revises plans for Okinawa, modifies Guam build-up. Retrieved from www.japantimes.co.jp/news/2014/04/19/national/politics-diplomacy/u-s-military-revises-plans-for-okinawa-modifies-guam-buildup/#.VbP1XPkYM90

Japan Times (2015, 25 November). Japan to deploy 500 GSDF troops on Ishigaki island. Retrieved from www.japantimes.co.jp/news/2015/11/25/national/politics-diplomacy/japan-deploy-500-gsdf-troops-ishigaki-island/#.VoJ-afkrLcc

Japanese Ant Image Database (2016). Okinawa-kuro-oo-ari. Retrieved from http://ant.edb.miyakyo-u.ac.jp/E/Taxo/F80905.html

JASSO (2016). *International students in Japan, 2015*. Japan Student Services Organisation. Retrieved from www.jasso.go.jp/en/about/statistics/intl_student/data2015.html

Jefferys, T. (1761). *The natural and civil history of the French dominions in North and South America. Part II*. London: T. Jefferys.

Jennings, K. (2014, 16 June). The borderless black hole in the middle of Europe. *Condé Nast Traveller*. Retrieved from www.cntraveler.com/stories/2014-06-16/lake-constance-maphead

Jiang, Jie (2014, 28 September). Xi clarifies Taiwan reunification position to visiting delegation. *Global Times*. Retrieved from www.globaltimes.cn/content/883880.shtml

Jin, Kai (2014, 30 April). The US, China, and the 'containment trap'. *The Diplomat*. Retrieved from http://thediplomat.com/2014/05/the-us-china-and-the-containment-trap/

JNTO (2016). Tourism statistics. Tokyo: Japan National Tourism Organisation. Retrieved from www.jnto.go.jp/jpn/statistics/tourists_2015.np.pdf (in Japanese).

Johnson, C. (2000). International Court of Justice-case concerning Kasikili/Sedudu island (Botswana/Namibia). *International Journal of Marine and Coastal Law*, *15*(4), 581–600.

Johnson, C. (2012). Toward post-sovereign environmental governance? Politics, scale and EU Water Framework Directive. *Water Alternatives*, *5*(1), 83–97.

Johnson, R. (2013, 12 January). China newspaper says to 'prepare for the worst' after military confrontation with Japan in the East China Sea. *Business Insider*. Retrieved from www.businessinsider.com/china-fighter-jet-flights-diaoyu-senkaku-islands-2013-1#ixzz3foC98Pn7

Joyner, C. C. (1987). The Antarctic minerals negotiating process. *The American Journal of International Law*, *81*(4), 888–905.

Joyner, C. C. (1992). *Antarctica and the Law of the Sea*. The Hague: Martinus Nijhoff.

Kabat, P., Bazelmans, J. van Dijk, J., Herman, P. M., van Oijen, T., Pejrup, M., Reise, K., Speelman, H. and Wolff, W. J. (2012). The Wadden Sea region: towards a science for sustainable development. *Ocean and Coastal Management*, *68*(1), 4–17.

Kacowicz, A. M. (1994). *Peaceful territorial change*. Columbia, SC: University of South Carolina Press.

Kaiman, J. (2014, 2 January). Chinese balloonist rescued by Japan near disputed islands. *Guardian*. Retrieved from www.theguardian.com/world/2014/jan/02/chinese-balloonist-disputed-islands-rescued-japan

Kakazu, H. (1994). *Sustainable development of small island economies*. Boulder, CO: Westview Press.

Kakazu, H. (2016). An island approach to the territorial issues in the South China Sea and the Senkaku/Diaoyu/Tiaoyutai islands. Paper presented at EWC/EWCA International Conference, Manila.

Kamata, S. (1983). *Japan in the passing lane: An insider's account of life in a Japanese auto factory*. New York: Random House.

Kang, D. C. (2005). Hierarchy in Asian international relations: 1300–1900. *Asian Security*, *1*(1), 53–79.

Kang, D. C. (2010). *East Asia before the West: Five centuries of trade and tribute*. New York: Columbia University Press.

Kaplan, R. D. (2014). *Asia's cauldron: The South China Sea and the end of a stable Pacific*. New York: Random House.

Kapur, K. D. (2012). Russia–Japan relations: politico-strategic importance of the disputed Southern Kurile Islands/Northern Territories. *India Quarterly: A Journal of International Affairs*, *68*(4), 385–405.

Karkkainen, B. C. (2004). Post-sovereign environmental governance. *Global Environmental Politics*, *4*(1), 72–96.

Kawashima, S. (2013, 5 June). The implications of the Japan–Taiwan fisheries agreement. Nippon.com. Retrieved from www.nippon.com/en/currents/d00081/

Kazianis, H. (2014, 1 July). China's 10 red lines in the South China Sea. *The Diplomat*. Retrieved from http://thediplomat.com/2014/07/chinas-10-red-lines-in-the-south-china-sea/

KBS World (2012, 5 October). Sightings of Chinese ships near Ieodo island increasing. Retrieved from http://world.kbs.co.kr/english/news/news_Po_detail.htm?No=93660&id=Po

Keane, M. (2006). From made in China to created in China. *International Journal of Cultural Studies*, *9*(3), 285–296.

Keck, Z. (2013, 27 November). US bombers challenge China's Air Defense Identification Zone. *The Diplomat*. Retrieved from http://thediplomat.com/2013/11/us-bombers-challenge-chinas-air-defense-identification-zone/

Keck, Z. (2015, 7 May). Watch Out, China: Japan Deploys 600 Troops, Missiles near Disputed Islands. *The National Interest*. Retrieved from www.nationalinterest.org/blog/the-buzz/watch-out-china-japan-ramps-military-forces-near-disputed-12831

Ker-Lindsay, J. (2014). Does a Cyprus solution still matter? In V. Lytra (ed.), *When Greeks and Turks meet: Interdisciplinary perspectives on the relationship since 1923* (pp. 109–124). Farnham: Ashgate.

Kerr, G. (2000). *Okinawa: The history of an island people*. North Clarendon, VT: Tuttle Publishing.

Kim, Y.-J. (2012, September 18). Why Ieodo matters. *Korea Times*. Retrieved from www.koreatimes.co.kr/www/news/nation/2012/09/117_120266.html

Kireev, G. V. (2006). *Rossiia–Kitai: Neizvestnye stranitsy pogranichnykh peregovorov* [Russia and China: Unknown pages of the boundary negotiations]. Moscow: Rosspen.

Kish, J. (1973). *The law of international spaces*. Leiden: Sijthoff.

Kliot, N. and Newman, D. (eds) (2013). *Geopolitics at the end of the Twentieth Century: The changing world political map*. London: Routledge.

Klostermann J. E. M. and Toonen, Th. A. J. (2011). *Besturen in het Waddengebied*. Working document. Leeuwarden: Waddenacademie (in Dutch).

Klostermann J. E. M., Revier, H., van den Berg, E. and Lamfers, J. (2009), *Je hebt nooit genoeg kennis. Onderzoek naar de kennisbehoefte in het Waddengebied*. Leeuwarden: Waddenacademie (in Dutch).

Knottnerus, O. S. (2005). History of human settlement, cultural change and interference with the marine environment. *Helgoland Marine Research*, *59*(1), 2–8.

Kodansha (1997). *Seclusion held: People, goods and information of Nagasaki Dejima*. Nagasaki: Kodansha Gendaishinsho.

Kotani, T. (2015, 20 October). The Japan–Taiwan fishery agreement: strategic success, tactical failure? *Asia Maritime Transparency Initiative*. Retrieved from https://amti.csis.org/the-japan-taiwan-fishery-agreement-strategic-success-tactical-failure/

Krasner, S. D. (1995). Compromising Westphalia. *International Security*, *20*(3), 115–151.

Kristiansen, R. (2013). Desolate dispute: a study of a hypothetical International Court of Justice (ICJ) decision. *Canadian Military Journal, 13*(3), 34–41. Retrieved from www.journal.forces.gc.ca/vol13/no3/doc/Kristiansen-Pages3441-eng.pdf

Kristof, N. D. (1999). 1492: The prequel. *New York Times Magazine*. Retrieved from http://partners.nytimes.com/library/magazine/millennium/m3/kristof.html

Kriwoken, L. K. and Holmes, N. (2007). Emerging issues of Australia's sub-Antarctic islands: Macquarie Island and Heard Island and McDonald Islands. In L. Kriwoken, J. Jabour and A. D. Hemmings (eds), *Looking South: Australia's Antarctic agenda* (pp. 149–164). Annandale, NSW: Federation Press.

Kriwoken, L. K., Jabour, J. and Hemmings, A. D. (2007). *Looking South: Australia's Antarctic agenda*. Sydney, NSW: Federation Press.

Kuo, Yu-Ying (2015). *Policy analysis in Taiwan*. Bristol: Policy Press at the University of Bristol.

La Parra Casado, D., Verdú, C. P. and Pérez, M. A. M. (2007). La imagen de España y Marruecos en la prensa marroquí y española durante el incidente del islote de Perejil (Leyla). *Revista CIDOB d'afers internacionals, 79–80*, 145–165 (in Spanish).

Lagoni, R. (1979). Oil and gas deposits across national frontiers. *American Journal of International Law*, *73*(2), 215–243.

Lake, J. (2000). *The republic of St. Martin*. Philipsburg, Sint Maarten: House of Nehesi.

Lamers, M., Eijgelaar, E. and Amelung, B. (2013). Last chance tourism in Antarctica. In R. H. Lemelin, J. Dawson and E. J. Stewart (eds), *Last chance tourism: Adapting tourism opportunities in a changing world* (pp. 25–41). London: Routledge.

Langer, W. L. (1951). *The diplomacy of imperialism: 1890–1902*. New York: Knopf.

Lapidus, G. W. (1987). The USSR and Asia in 1986: Gorbachev's new initiatives. *Asian Survey*, *27*(1), 1–9.

Laracy, H. (ed.) (1983). *Tuvalu: A history*. Suva: Institute of Pacific Studies, University of the South Pacific.

Lauterpacht, E. (1970). *International law, being the collected papers of Hersch Lauterpacht. Systematically arranged and edited by E. Lauterpacht*. Cambridge: Cambridge University Press.

Lavelle, M. and Smith, J. (2012, 26 October). Why are China and Japan sparring over eight tiny, uninhabited islands? *National Geographic*. Retrieved from http://news.nationalgeographic.com/news/energy/2012/10/121026-east-china-sea-dispute/

Lawrence, D. H. (1928/1986). *The man who loved islands*. Lewisburg, PA: Ellen Clarke Bertrand Library, Bucknell University.

Leb, C. (2013). *Cooperation in the law of transboundary water resources*. Cambridge: Cambridge University Press.

Lebow, R. N. (1983). Miscalculation in the South Atlantic: the origins of the Falkland war. *The Journal of Strategic Studies*, *6*(1), 5–35.

Lee, H. (2016). Cruise tourism in Northeast Asia. In A. Papathanasis (ed.), *Cruise business development: Safety, product design and human capital* (pp. 81–93). Heidelberg: Springer International Publishing.

Lee, I. and Ming, F. (2012). Deconstructing Japan's claim of sovereignty over the Diaoyu/Senkaku islands. *The Asia-Pacific Journal, 10* (Issue 53, No. 1), 1–23. Retrieved from www.japanfocus.org/-Fang-Ming/3877/article.html

Lee, J. (2011). Senkaku/Diaoyu: islands of conflict. *History Today*, *61*(5). Retrieved from www.historytoday.com/joyman-lee/senkakudiaoyu-islands-conflict

Lee, J. T. (2015). Chinese innovations in sovereignty: *Neidi*, frontiers and borders. Irvine CA: University of California at Irvine. Retrieved from www.omnifoo.info/images/Chinese%20Innovations%20in%20Sovereignty.pdf

Lee, Mingjiang and Shoon, Ming Hui (2015). From cooperation for competition to non-confrontational assertiveness: China's changing responses to US regional policies in the Asia-Pacific. In Y. Sato and S. S. Tan (eds), *United States engagement in the Asia Pacific: Perspectives from Asia* (pp. 214–224). Amherst, NY: Cambria Press.

Lee, S. (2002). The 1951 San Francisco Peace Treaty with Japan and the territorial disputes in East Asia. *Pacific Rim Law & Policy Journal*, *11*(1), 65–144. Retrieved from https://digital.law.washington.edu/dspace-law/bitstream/handle/1773.1/748/11PacRimLPolyJ063.pdf?sequence=1

Lerhe, E. (2015). *The Asia-Pacific and the Royal Canadian Navy*. Canadian Global Affairs Institute. June. Retrieved from www.cgai.ca/the_asia_pacific_and_the_royal_canadian_navy#2011

Levenstein, S. (2011). Shared island between China and Russia to become eco-tourist zone. *China News*. July. Retrieved from http://inventorspot.com/articles/shared_island_become_ecotourist_zone_china_and_russia

Lieberthal, K. (2011, 21 December). The American pivot to Asia. *Foreign Policy*, *21*. Retrieved from https://members.marshallcenter.org/ar/system/files/The%20American%20Pivot%20to%20Asia%20Lieberthal.pdf

Lin, Man-houng (2014, 26 May). Taiwan and the Ryukyus (Okinawa) in Asia-Pacific multilateral relations. *The Asia-Pacific Journal: Japan Focus*, *11*(21), 3. Retrieved from www.japanfocus.org/-Lin-Man_houng/4121/article.html

Lindström, B. (2015). *Beskattningen och den åländska självstyrelsens finansiering: Dagsläge och historisk bakgrund.* Stockholm: YMER 2015. Svenska sällskapet för Antropologi och Geografi (in Swedish).

Liu, Ming (2015). Northeast Asia Order after World War II: continuity, compliance, power-transition and challenges. *The Korean Journal of Defence Analysis*, *27*(2), 163–186.

Logan, J. (2013, 8 January). China, America, and the pivot to Asia. *Cato Institute Policy Analysis*, *717*. Retrieved from http://papers.ssrn.com/sol3/Delivery.cfm/SSRN_ID2228171_code1953816.pdf?abstractid=2228171&mirid=1

Loope, L. L., Hamann, O. and Stone, C. P. (1988). Comparative conservation biology of oceanic archipelagoes: Hawaii and the Galapagos. *BioScience*, *38*(4), 272–282.

Love (2015). *Royalty in blood.* TV series. Beijing. Retrieved from http://cn.lovetvshow.info/2015/07/royalty-in-blood-list.html

Lovell, J. (2011). *The opium war: Drugs, dreams and the making of China*. Sydney, NSW: Picador.

Luiz, M. T. and Schillat, M. (1998). *Tierra del Fuego: Materiales para el estudio de la historia regional*. Ushuaia: Editorial Fuegia (in Spanish).

Lutz, C. and Enloe, C. (eds) (2009). *The bases of empire: The global struggle against US military posts*. New York: NYU Press.

Macdonald, B. (1982). *Cinderellas of the empire: Towards a history of Kiribati and Tuvalu*, Canberra: Australian National University Press.

Macdonald, B. (1975). Secession in the defence of identity. *Pacific Viewpoint*, *16*(1), 26–44.

MacDonald, F. (2006). The last outpost of Empire: Rockall and the Cold War. *Journal of Historical Geography*, *32*(3), 627–647.

Macfarlane, R. (2012). *The old ways: A journey on foot*. London: Penguin.

MacFarquhar, R., Fairbank, J. K. and Twitchett, D. C. (1991). *The Cambridge history of China*. Cambridge: Cambridge University Press.

Mackinder, H. J. (1904). The geographical pivot of history. *The Geographical Society*, *23*(4), 421–437.

Mackrael, K. (2012, 29 November). Canada, Denmark closer to settling border dispute. *The Globe and Mail*. Retrieved from www.theglobeandmail.com/news/national/canada-denmark-closer-to-settling-border-dispute/article5831571/?page=all

Maluwa, T. (1993). Disputed sovereignty over Sidudu (or Kasikili) island (Botswana–Namibia): some observations on the international legal aspects. *African Journal of International & Comparative Law*, *5*(1), 113–138.

Mao, Feng (2010, 24 September). Japan releases captain after China stops rare earth metal exports. Translated by Vivian Po. *New America Media*. Retrieved from http://newamericamedia.org/2010/09/japan-releases-captain-after-china-stops-rare-earth-metal-exports.php

Marquand, R. (2005, 11 April). Anti-Japan protests jar an uneasy Asia: demonstrations spread from Beijing to several southern cities Sunday. *Christian Science Monitor*. Retrieved from www.csmonitor.com/2005/0411/p01s04-woap.html

Martin, R. (2012, 12 July). Apple pulls controversial Chinese Island defense game from App store. *TechinAsia*. Retrieved from www.techinasia.com/apple-diaoyu-island-app-game-pulled-senkaku

Maslow, S., Mason, R. and O'Shea, P. (2015). *Risk state: Japan's foreign policy in an age of uncertainty*. Farnham: Ashgate.

Maude, H. E. (1968). *Of islands and men: Studies in Pacific history*, Melbourne, VIC: Oxford University Press.

Maxwell, N. (2007). How the Sino-Russian boundary conflict was finally settled: from Nerchinsk 1689 to Vladivostok 2005 via Zhenbao Island 1969. *Critical Asian Studies*, *39*(2), 229–253.

May, R. J., Matbob, P. and Papoutsaki, E. (2013). New Guinea. In G. Baldacchino (ed.), *The political economy of divided islands: Unified geographies, multiple polities* (pp. 34–58). London: Palgrave Macmillan.

McCormack, G. and Norimatsu, S. O. (2012). *Resistant islands: Okinawa confronts Japan and the United States*. Lanham, MD: Rowman & Littlefield.

McCourt, D. M. (2011). Role-playing and identity formation in international politics: Britain's reinvasion of the Falklands, 1982. *Review of International Studies*, *37*(4), 1599–1621.

McCullough, J. J. (2004, 31 March). Hans Off! Retrieved from www.jjmccullough.com/index.php/2004/03/31/hans-off/

McElroy, J. L. and De Albuquerque, K. (1998). Tourism penetration index in small Caribbean islands. *Annals of Tourism Research*, *25*(1), 145–168.

McIntyre, W. D. (2012). The partition of the Gilbert and Ellice Islands. *Island Studies Journal*, *7*(1), 135–146.

McIntyre, W. D. (2014). *Winding up the British Empire in the Pacific Islands*. Oxford: Oxford University Press.

McKean, M. and Ostrom, E. (1995). Common property regimes in the forest: just a relic from the past. *Unasylva*, *46*(180), 3–15.

Megaw, M. R. (1977). The scramble for the Pacific: Anglo-United States rivalry in the 1930s. *Australian Historical Studies*, *17*(69), 458–473.

Meierding, E. (2015, 27 May). Don't blame the oil rigs. *Pacific Standard*. Retrieved from www.psmag.com/politics-and-law/dont-blame-the-oil-rigs

Menzies, G. (2003). *1421: The year China discovered the world*. New York: Random House.

Miller, C. A. and Edwards, P. N. (2001). *Changing the atmosphere: Expert knowledge and environmental governance*. Cambridge, MA: MIT Press.

Milmo, C. (2016, 3 January). Bitcoin: how the Isle of Man is leading a cryptocurrency revolution. *Independent.* Retrieved from www.independent.co.uk/news/uk/home-news/bitcoin-how-the-isle-of-man-is-sparking-a-cryptocurrency-revolution-a6794756.html

Milun, K. (2011). *The political uncommons: The cross-cultural logic of the global commons*. Farnham: Ashgate.

Ministry of Foreign Affairs, Japan (2014). *Territory of Japan*. Retrieved from www.mofa.go.jp/territory/index.html

Ministry of Foreign Affairs, Japan (2015). *About the Senkaku Islands*. Retrieved from www.mofa.go.jp/region/asia-paci/senkaku/

Ministry of Foreign Affairs, Japan (2016a). Japanese territory: Senkaku islands. Retrieved from www.mofa.go.jp/region/asia-paci/senkaku/

Ministry of Foreign Affairs, Japan (2016b). *The Senkaku Islands: Seeking maritime peace based on the rule of law, not force or coercion*. Retrieved from www.mofa.go.jp/files/000018519.pdf

Ministry of Foreign Affairs of the PRC (2001, 24 July). *Treaty of good-neighbourliness and friendly cooperation between the People's Republic of China and the Russian Federation*. Retrieved from www.fmprc.gov.cn/eng/wjdt/2649/t15771.htm

Ministry of Foreign Affairs of the PRC (2014). *Set aside dispute and pursue joint development*. Retrieved from www.fmprc.gov.cn/mfa_eng/ziliao_665539/3602_665543/3604_665547/t18023.shtml

Ministry of Foreign Affairs of the PRC (2015). The Department of Hong Kong, Macao and Taiwan Affairs. Retrieved from www.fmprc.gov.cn/mfa_eng/wjb_663304/zzjg_663340/gats_665294/

Ministry of Foreign Affairs, ROC (2012, 15 August). *East China Sea Peace Initiative*. Retrieved from www.mofa.gov.tw/en/theme.aspx?s=780E70E6D142B833&sms=BCDE19B435833080

Miyoshi, M. (1999). *The joint development of offshore oil and gas in relation to maritime boundary delimitation*. Durham: International Boundaries Research Unit.

Miyoshi, M. (2014, 17 October). Exercising enforcement jurisdiction around the Senkaku islands. *Review of Island Studies*. Retrieved from http://islandstudies.oprf-info.org/wp/wp-content/uploads/2014/12/a00013r.pdf

Montreal Gazette (2016, 17 February). Fairytale island home of Céline Dion has been sold. Retrieved from http://montrealgazette.com/business/local-business/real-estate/celine-dions-laval-mansion-sells-for-25-5-million-reports-say

Moran, J. and Rottman, G. L. (2002). *Peleliu 1944*. Oxford: Osprey Publishing.

Morgan, G. (1996). *Images of organisation*. London: Sage.

Moss, L. (2014, 18 September). Wildlife thrives in 'the most dangerous place on Earth'. *Mother Nature Network*. Retrieved from www.mnn.com/lifestyle/eco-tourism/stories/wildlife-thrives-in-the-most-dangerous-place-on-earth

Mottola, P. T. (1999). Article III, Section 2, Clause 2. Original Jurisdiction of the US Supreme Court: in an interstate boundary dispute, New Jersey was granted sovereign authority over the landfilled portion of Ellis Island, with New York retaining sovereign authority over the original portion, pursuant to the express terms of the States' Compact of 1834: New Jersey v. New York, 118 S. Ct. 1726 (1998). *Seton Hall Constitutional Law Journal*, *9*, 1113–1195.

Mountz, A. (2011). The enforcement archipelago: detention, haunting, and asylum on islands. *Political Geography*, *30*(1), 118–128.

Munro, D. (1996). Samoan pastors in Tuvalu, 1865–1899. In D. Munro and A. Thornley (eds), *The covenant makers: Islander missionaries in the Pacific* (pp. 124–157). Suva: Pacific Theological College and the University of the South Pacific.

Munro, D. and Firth, S. (1986). Towards colonial protectorates: the case of the Gilbert and Ellice islands. *Australian Journal of Politics and History*, *32*(1), 63–71.

Murphy, K. (2014, 12 August). Australia embraces missile and naval ties as it cements US defence pact. *Guardian*. Retrieved from www.theguardian.com/world/2014/aug/12/australia-embraces-missile-naval-ties-us-defence-pact

Murphy, R. T. (2014). *Japan and the shackles of the past*. Oxford: Oxford University Press.

Myers, R. H. (1996). A new Chinese civilisation: the evolution of the Republic of China on Taiwan. *The China Quarterly*, *148*, 1072–1090.

Nagata, K. (2011, 30 January). Conservatives rally against DPJ. *Japan Times*. Retrieved from www.japantimes.co.jp/news/2011/01/30/national/conservatives-rally-against-dpj/#.V4IxyfkrLIU

Nakashima, K. (2012, 17 April). Governor Ishihara: 'Tokyo is going to protect the Senkaku islands. Any complaints?' Retrieved from https://jsmyth.wordpress.com/2012/04/18/tokyo-senkaku/ (original in Japanese retrieved from Yomiuri Shimbun).

Nakasone, R. Y. (2002). *Okinawan diaspora*. Honolulu, HI: University of Hawai'i Press.

National Post (2012, 11 April). New proposal would see Hans Island split equally between Canada and Denmark. Retrieved from http://news.nationalpost.com/2012/04/11/new-proposal-would-see-hans-island-split-equally-between-canada-and-denmark/

Nester, W. (1993). Japan, Russia and the Northern Territories: continuities, changes, obstacles, opportunities. *Third World Quarterly*, *14*(4), 717–734.

Neuman, W. and Ayala, M. (2012, 16 August). Ecuador grants asylum to Assange, defying Britain. *New York Times*. Retrieved from www.nytimes.com/2012/08/17/world/americas/ecuador-to-let-assange-stay-in-its-embassy.html?_r=0

New York Times Editorial (2014, 2 March). Mr. Abe's dangerous revisionism. *New York Times*. Retrieved from www.nytimes.com/2014/03/03/opinion/mr-abes-dangerous-revisionism.html?_r=0

Nickson, R. A. (1982). The Itaipú hydro-electric project: the Paraguayan perspective. *Bulletin of Latin American Research, 2(1), 1–20. Retrieved from* www.jstor.org/stable/3338386

Nie, H. A. (2013). Gaming, nationalism, and ideological work in contemporary China: online games based on the war of resistance against Japan. *Journal of Contemporary China*, *22*(81), 499–517.

Numminen, L. (2011). A history and functioning of the Spitsbergen Treaty. In D. Wallis and S. Arnold (eds), *The Spitsbergen Treaty: Multilateral governance in the Arctic* (pp. 7–20). Helsinki: Arctic Papers 1.

O'Connell, D. P. (1970). Condominium of the New Hebrides. In *The British Yearbook of International Law, 1968–1969* (pp. 71–46). London: Oxford University Press.

O'Connor, S. and Hardy, J. (2015, 15 February). Imagery shows progress of Chinese land building across Spratlys. *IHS Jane's 360*. Retrieved from www.janes.com/article/48984/imagery-shows-progress-of-chinese-land-building-across-spratlys

Obama, Barack (2014, 24 April). Joint Press Conference with President Obama and Prime Minister Abe of Japan, Akasaka Palace, Tokyo, Japan. Washington, DC: The White House. Retrieved from www.whitehouse.gov/the-press-office/2014/04/24/joint-press-conference-president-obama-and-prime-minister-abe-japan

Obe, M. (2014, 19 August). Japan–China trade set to grow first time in three years. *Wall Street Journal*. Retrieved from www.wsj.com/articles/japan-china-trade-set-to-grow-first-time-in-3-years-1408448464

Olson, S. L. and Hearty, P. J. (2003). Probable extirpation of a breeding colony of short-tailed albatross (*Phoebastria albatrus*) on Bermuda by Pleistocene sea-level rise. *Proceedings of the National Academy of Sciences, 100*, 12825–12829.

Oosterhaven, J. (1981). *Interregional input–output analysis and Dutch regional policy problems*. Aldershot: Gower.

Ostrom, E. (1990). *Governing the commons: The evolution of institutions for collective action*. Cambridge: Cambridge University Press.

Otani, S., Ohno, G., Shimoeda, N. and Mikami, H. (2004). Morbidity and health survey of wintering members in Japanese Antarctic research expedition. *Circumpolar Health, 63*(Suppl. 2), 165–168.

Page, T., Hannington, T., Bunt, C., Potrawiak, A. and Berry, A. (2012). *Opportunities for the smallholder sandalwood industry in Vanuatu*. Technical Report No. 79. Canberra, Australia: Australian Centre for International Agricultural Research. Retrieved from http://researchonline.jcu.edu.au/23325/1/23325_Page_et_al_2012.pdf

Page, T., Potrawiak, A., Berry, A., Tate, H., Tungon, J. and Tabi, M. (2010). Production of sandalwood (*Santalum austrocaledonicum*) for improved smallholder incomes in Vanuatu. *Forests, Trees and Livelihoods, 19*(3), 299–316.

Paichadze, S. and Seaton, P. A. (eds) (2015). *Voices from the shifting Russo-Japanese border: Karafuto/Sakhalin*. London: Routledge.

Palan, R. (2006). *The offshore world: Sovereign markets, virtual places and nomad millionaires*. Ithaca, NY: Cornell University Press.

Palonkorpi, M. (ed.) (2015). *The South Caucasus beyond borders, boundaries and division lines: Conflicts, cooperation and development*. Turku: Ministry for Foreign Affairs of Finland and Aleksanteri Institute, University of Finland.

Panda, A. (2015, 14 January). Japan approves largest-ever defence budget. *The Diplomat*. Retrieved from http://thediplomat.com/2015/01/japan-approves-largest-ever-defense-budget/

Pao-Min, C. (1990). A new scramble for the South China Sea islands. *Contemporary Southeast Asia, 12*(1), 20–39.

Patrikeeff, F. (2002). *Russian politics in exile: The northeast Asian balance of power, 1924–1931*. New York: Palgrave Macmillan.

Pauly, D. (2009). The fisheries resources of the Clipperton Island EEZ (France). *Fisheries Centre Research Reports, 17*, 35–37. Retrieved from http://aerl06.aerl.ubc.ca/doc/CatchReconstruction/EEZ/Clipperton-Pauly-2009.pdf

PEP (1998). *Protocol on environmental protection to the Antarctic Treaty*. Retrieved from www.ats.aq/documents/recatt/Att006_e.pdf

Perdue, P. C. (2011). *The First Opium War*. Cambridge, MA: Massachusetts Institute of Technology Visualising Cultures Programme. Retrieved from http://ocw.mit.edu/ans7870/21f/21f.027/opium_wars_01/ow1_essay01.html

Perkins, T. C. (2014). Edification from the Andorran model: a brief exploration into the condominium solution on the international stage and its potential application to current land disputes. *Indiana Journal of Global Legal Studies, 21*(2), 643–665.

Perlez, J. (2014, 7 November). China and Japan, in sign of a thaw, agree to disagree on a disputed island group. *New York Times*. Retrieved from www.nytimes.com/2014/11/08/world/asia/china-japan-reach-accord-on-disputed-islands-senkaku-diaoyu.html?_r=0

Perlez, J. (2015, 16 April). China building aircraft runway in disputed Spratly islands *New York Times*. Retrieved from www.nytimes.com/2015/04/17/world/asia/china-building-airstrip-in-disputed-spratly-islands-satellite-images-show.html?_r=0

Permanent Court of Arbitration (2016, 12 July). The South China Sea Arbitration (The Republic of the Philippines v. The People's Republic of China). Retrieved from https://pca-cpa.org/wp-content/uploads/sites/175/2016/07/PH-CN-20160712-Press-Release-No-11-English.pdf

Permanent Court of International Justice (1933, 5 September). Legal status of Eastern Greenland: Denmark v. Norway Judgment. Twenty-Sixth Session. Retrieved from www.worldcourts.com/pcij/eng/decisions/1933.04.05_greenland.htm

Perry, N. (2011, 27 October). Tuvalu's fresh water crisis deepens: contamination of groundwater, scarce rain, threaten way of life. *The Washington Times*. Retrieved from www.washingtontimes.com/news/2011/oct/27/tuvalus-fresh-water-crisis-deepens/?page=all

Pinilla, D. G. (2004). Border disputes between China and North Korea. *China Perspectives*. Retrieved from: http://chinaperspectives.revues.org/806

Podder, S. (2012). Legitimacy, loyalty and civilian support for the Moro Islamic Liberation Front: changing dynamics in Mindanao, Philippines. *Politics, Religion & Ideology*, *13*(4), 495–512.

Posner, E. (2014, 25 February). Why are China and Japan inching toward war over five tiny islands? *Slate Magazine*. Retrieved from www.slate.com/articles/news_and_politics/view_from_chicago/2014/02/the_senkaku_or_diaoyu_islands_where_world_war_iii_could_start_because_of.html

Post, H. H. (2016). Dutch–German boundary relations in the Eems-Dollard (Ems-Dollart) estuary: an implicit condominium? In M. Nicolini, F. Palermo and E. Milano (eds), *Law, territory and conflict resolution: Law as a problem and law as a solution* (pp. 346–361). Leiden: E. J. Brill.

Potsdam Declaration (1945). Retrieved from www.ndl.go.jp/constitution/e/etc/c06.html

Powell, E. (2013). Taming the beast: how the international legal regime creates and contains flags of convenience. *Annual Survey of International & Comparative Law 19*(1), Article 12. Retrieved from http://digitalcommons.law.ggu.edu/annlsurvey/vol19/iss1/12

Prasad, N. (2004). Escaping regulation, escaping convention: development strategies in small economies. *World Economics*, *5*(1), 41–65.

Prescott, J. R. V. (1986[1975]). *The maritime political boundaries of the world*. London: Methuen.

Press, S. M. (2010). To govern, or not to govern: Prussia, neutral Moresnet. *Social Science Research Network*. Retrieved from http://ssrn.com/abstract=2096313

PressTV (2013, 12 March). British MP: UK, Argentina should share sovereignty over Falklands. Retrieved from www.presstv.com/detail/2013/03/12/293182/falklands/

Przystup, J. J. and Saunders, P. C. (2013, 27 February). Time for China and Japan to cool it. *Yale Global On-line*. Retrieved from http://yaleglobal.yale.edu/content/time-china-and-japan-cool-it

Puente-Rodríguez, D., Giebels, D. and de Jonge, V. N. (2015). Strengthening coastal zone management in the Wadden Sea by applying 'knowledge-practice interfaces'. *Ocean and Coastal Management*, *108*(1), 27–38.

Pyle, K. B. (2007). *Japan rising: The resurgence of Japanese power and purpose*. New York: Public Affairs.

Qiu, Jin (2006). The politics of history and historical memory in China–Japan relations. *Journal of Chinese Political Science*, *11*(1), 25–53.

Ramos-Mrosovsky, C. (2009) International law's unhelpful role in the Senkaku islands. *University of Pennsylvania Journal of International Law*, *29*(4): 903–946.

Rawlings, G. (2012). Statelessness, human rights and decolonisation: citizenship in Vanuatu, 1906–80. *The Journal of Pacific History*, *47*(1), 45–68.

Redfield, P. (2000). *Space in the tropics: From convicts to rockets in French Guiana.* Berkeley, CA: University of California Press.

Refsing, K. (2003). In Japan but not of Japan. In C. Mackerras (ed.) *Ethnicity in Asia* (pp. 48-63). New York: Routledge.

Rehman, I. (2014, 28 February). Why Taiwan matters: a small island of great strategic importance. *The National Interest.* Retrieved from http://nationalinterest.org/commentary/why-taiwan-matters-9971?page=2

Reise, K. (2005). Coast of change: habitat loss and transformations in the Wadden Sea. *Helgoland Marine Research*, *59*(1), 9–21.

Reise, K. (2011). Das wattenmeer: wirklichkeiten und visionen. In L. Fischer and K. Reise (eds), *Küstenmentalität und Klimawandel* (pp. 167–179). Munich: Oekum.

Reise, K. (2013). *A natural history of the Wadden Sea: Riddled by contingencies.* Leeuwarden and Wilhemshaven. Wadden Academy & CWSS.

Reise, K., Baptist, M., Burbridge, P., Dankers, N., Fischer, L., Flemming, B., Oost, A. P. and Smit, C. (2010). The Wadden Sea: a universally outstanding tidal wetland. Wadden Sea Ecosystem No. 29. Common Wadden Sea Secretariat, Wilhelmshaven.

Rekenkamer (2013). *Waddengebied: natuurbescherming, natuurbeheer en ruimtelijke inrichting.* The Hague: Algemene Rekenkamer (in Dutch).

Reuters (2014, 9 April). China angered by latest U.S. arms sale plan for Taiwan. Retrieved from www.reuters.com/article/2014/04/09/us-china-usa-taiwan-idUSBREA380H520140409

Reynolds, R. (2005, 12 August). Who owns Hans Island? *NPR News.* National Public Radio. Web. Transcript.

Rezvani, D. A. (2014). *Surpassing the sovereign state: The wealth, self-rule and security advantages of partially independent territories.* Oxford: Oxford University Press.

Rezvani, D. A. (2015). Partial independence beats full independence. *Territory, Politics, Governance*, *4*(3), 269–296.

Riabushkin, D. S. (2006). What they fought with on Damanskii island. *Journal of Slavic Military Studies*, *19*(1), 149–166.

Richardson, P. B. (2016). Beyond the nation and into the state: identity, belonging, and the 'hyper-border'. *Transactions of the Institute of British Geographers*, *41*(2), 201–215.

Rivera-Batiz, F. and Santiago, C. E. (1996). *Island paradox: Puerto Rico in the 1990s.* New York: Russell Sage Foundation.

Robson, R. W. (1963). New Hebrides. In *Pacific islands yearbook and who's who* (pp. 431–444). 9th edn. Sydney, NSW: Pacific Publications.

Ronström, O. (2013). Finding their place: islands as locus and focus. *Cultural Geographies*, *20*(2), 153–165.

Rosie, M. (2014). Scotland's referendum. *Scottish Affairs*, *23*(3), 275–279.

Ross, R. S. (2002). Navigating the Taiwan Strait: deterrence, escalation dominance and U.S. –China relations. *International Security*, *27*(2), 48–85.

Royle, S. A. (2001). *A geography of islands: Small island insularity.* London: Routledge.

Royle, S. A. (2011). Conclusion: lessons from islands, or islands as miner's canaries. In G. Baldacchino (ed.), *Extreme heritage management: The practices and policies of densely populated islands* (pp. 246–259). New York: Berghahn.

RT (2008, 14 October). World's biggest country becomes a little bit smaller. Retrieved from www.rt.com/news/worlds-biggest-country-becomes-a-little-bit-smaller/

Runhaar, H., Runhaar, P. R. and Oegema, T. (2010). Food for thought: conditions for discourse reflection in the light of environmental assessment. *Environmental Impact Assessment Review*, *30*(6), 339–346.

Ruppel, O. C. (2015). Climate change, law and development in Africa: a reflection on selected aspects, relations and responses. In H.-J. Koch, D. König, J. Sanden and R. Verheyen (eds), *Legal regimes for environmental protection: Governance for climate change and ocean resources* (pp. 89–120). Leiden: Brill.

Ryall, J. (2014, 11 September). Japan extends its sea territory by 177,000 sq km in line with UN ruling. *South China Morning Post*. Retrieved from www.scmp.com/news/asia/article/1589572/japan-extends-its-sea-territory-177000-sq-km-line-un-ruling

Ryukyu Shimpo (2012, 20 September). More than 3,000 Chinese cancel travel to Okinawa because of souring relations between Japan and China. Retrieved from http://english.ryukyushimpo.jp/2012/09/29/8189/

Sakai, R. K. (1964). The Satsuma–Ryukyu trade and the Tokugawa seclusion policy. *The Journal of Asian Studies*, *23*(3), 391–403.

Salazar, J. F. (2015, 9 August). Antarctica may hold the key to regulating mining in space. *The Conversation*. Retrieved from http://theconversation.com/antarctica-may-hold-the-key-to-regulating-mining-in-space-45702

Salman, S. M. (2000). International rivers as boundaries: the dispute over Kasikili/Sedudu island and the decision of the International Court of Justice. *Water International*, *25*(4), 580–585.

Samuels, J. H. (2007). Condominium arrangements in international practice: reviving an abandoned concept of boundary dispute resolution. *Michigan Journal of International Law*, *29*, 727–775.

Samuels, M. (2013). *Contest for the South China Sea*. New York: Routledge.

San Francisco Peace Treaty (1951). Retrieved from http://archive.alpha-canada.org/SFPTGA/SanFranciscoPeaceTreaty1951.htm

Scarpulla, C. (1999). *The constitutional framework for the autonomy of Åland: A survey of the status of an autonomous region in the throes of European integration*. Mariehamn: Ålands Högskola.

Schlicher, K. (2012, 1 January). Land and luxury: celebrities who have their own island. *Madame Noire*. Retrieved from http://madamenoire.com/123448/land-luxury-celebrities-who-have-their-own-island/

Schneider, J. (1992). Condominium. In R. Bernhardt (ed.), *Encyclopaedia of public international law: Vol. 1* (pp. 732–734). Amsterdam: Elsevier Science.

Schoenbaum, T. J. (2008). Introduction: resolving the disputes on the basis of peace, justice and international cooperation. In T. J. Schoenbaum (ed.), *Peace in Northeast Asia: Resolving Japan's territorial and maritime disputes* (pp. 1–6). Cheltenham: Edward Elgar.

Schofield, C. H. (2005). Dividing the resources of the Timor Sea: a matter of life and death for East Timor. *Contemporary Southeast Asia*, *27*(2), 255–280.

Schofield, C. H. (2007). Minding the gap: the Australia–East Timor Treaty on certain maritime arrangements in the Timor Sea (CMATS). *International Journal of Marine and Coastal Law*, *22*(2), 189–234.

Schofield, C. H. (2014). Defining areas for joint development in disputed waters. In S. Wu and N. Hong (eds), *Recent developments in the South China Sea and prospects for joint development* (pp. 78–98). New York: Routledge.

Schofield, C. H. and Wang, D. K.-H. (2012). The regime of islands under UNCLOS: implications for the South China Sea. In C. Schofield (ed.), *Maritime energy resources*

in Asia: Legal regimes and cooperation (pp. 61–77). Seattle, WA: National Bureau of Asian Research.
Schulte, C. (2004). *Compliance with decisions of the International Court of Justice*. Oxford: Oxford University Press.
Schuman, M. (2013, 1 December). China and Japan may not like each other, but they need each other. *Time*. Retrieved from http://world.time.com/2013/12/01/china-and-japan-may-not-like-each-other-but-they-need-each-other/
Scobell, A. (2000). Show of force: Chinese soldiers, statesmen, and the 1995–1996 Taiwan Strait crisis. *Political Science Quarterly*, *115*(2), 227–246.
Scott, S. V. (1999). Japan's renunciation of territorial rights in Antarctica and Australian diplomacy. *Polar Record*, *35*(193), 99–106.
Scoville, R. M. (2014). A defence of Japanese sovereignty over the Senkaku/Diaoyu islands. *George Washington International Law Review*, *46*(3), 13–16.
Scoville, R. M. (2015). Sovereignty over the Senkaku/Diaoyu islands: a guide to the legal debate. In T. F. Liao, K. Hara and K. Wiegand (eds), *The China–Japan border dispute: Islands of contention in multidisciplinary perspective* (pp. 85–114). Farnham: Ashgate.
Secretariat of the Atlantic Treaty (2015). *The Antarctic Treaty*. Retrieved from www.ats.aq/documents/ats/treaty_original.pdf
Sehdev, P. S. (2002). The origin of quarantine. *Clinical Infectious Diseases*, *35*(9), 1071–1072.
Seidl-Hohenveldern, I. (1988). Extraterritorial respect for state acts. *Hague Yearbook of International Law*. Hague Academy of International Law. Association of Attenders and Alumni. The Hague: Martinus Nijhoff.
Sekou, L. M., Francis, O. and Gumbs, N. (eds) (1990). *The Independence papers, vol. I: Readings on a new political status for St. Maarten/St. Martin*. Philipsburg, Sint Maarten: House of Nehesi.
Selden, M. and Susumu, Y. (2014). The origins of the Senkaku/Diaoyu dispute between China, Taiwan and Japan. *Asia-Pacific Journal: Japan Focus*, *12*(2), No. 3. Retrieved from http://japanfocus.org/-Mark-Selden/4061/article.html
Shah, S. K. (2015). *India and China: The battle between soft and hard power*. New Delhi: Vij Books India.
Shambaugh, D. (2013). *China goes global: The partial power*. New York: Oxford University Press.
Shanghai Daily (2016, 30 June). China says arbitral tribunal has no jurisdiction over South China Sea case. Retrieved from www.shanghaidaily.com/article/article_xinhua.aspx?id=330337
Sharkey, H. J. (2003). *Living with colonialism: Nationalism and culture in the Anglo-Egyptian Sudan*. Berkeley, CA: University of California Press.
Shaw, Han-Yi (1999). *The Diaoyutai/Senkaku islands dispute: Its history and an analysis of the ownership claims of the P.R.C., R.O.C., and Japan*. College Park, MD: University of Maryland., Occasional Papers/Reprint Series, No. 152.
Shaw, Han-Yi (2012, 19 September). The inconvenient truth behind the Diaoyu/Senkaku islands, *New York Times*, Retrieved from http://blog.hiddenharmonies.org/2012/09/19/the-inconvenient-truth-behind-the-diaoyusenkaku-islands-by-han-yi-shaw/
Shaw, M. N. and Evans, M. D. (2000). Case concerning Kasikili/Sedudu island (Botswana/Namibia). *International and Comparative Law Quarterly*, *49*(4), 964–978.
Shelton, M. (1945, 5 April). Ryukyuans have 'double allegiance' – Schwartz. *The Stanford Daily*, *107*(44). Retrieved from http://stanforddailyarchive.com/cgi-bin/stanford?a=d&d=stanford19450405-01.2.19#

Shen, Wenwen (2012, 1 March). China and its neighbours: troubled relations. *EU–Asia Centre*. Retrieved from www.eu-asiacentre.eu/pub_details.php?pub_id=46

Sheng-Ping, H. (2014). General assessment of the PRC's foreign policy: new initiatives and orientations under Xi Jinping. *Prospect Journal, 11*(1), 67–98.

Shibuichi, D. (2005). The Yasukuni Shrine dispute and the politics of identity in Japan: why all the fuss? *Asian Survey*, *45*(2), 197–215.

Shih, Hsiu-chuan (2013, 11 April). Taiwan, Japan ink fisheries agreement. *Taipei Times*. Retrieved from www.taipeitimes.com/News/front/archives/2013/04/11/2003559323

Shine, C. (1997). Legal mechanisms to strengthen and safeguard trans-boundary protected areas. In *Parks for Peace Conference Proceedings* (pp. 37–47). Somerset: Peace Parks Foundation, 16–18 September.

Short, G. and Carrington, B. (1996). Anti-racist education, multiculturalism and the new racism, *Educational Review, 48*(1), 65–77.

Sicking, L. (2006). *Frontières d'outre-mer: La France et les Pays-Bas dans le monde atlantique au XIXe siècle*. Paris, France: Les Indes Savantes (in French).

Sidaway, J. D. (2007). Spaces of postdevelopment. *Progress in Human Geography, 31*(3), 345–361.

Sijtsma, F. J., Daams, M. N., Farjon, H. and Buijs, A. E. (2012). Deep feelings around a shallow coast: a spatial analysis of tourism jobs and the attractivity of nature in the Dutch Wadden area. *Ocean and Coastal Management, 68*(1), 128–138.

Silk, N. and Ciruna, N. (2013). *A practitioner's guide to freshwater biodiversity conservation*. Washington, DC: Island Press.

Silverstein, A. L. (1990). Okinotorishima: artificial preservation of a speck of sovereignty. *Brooklyn Journal of International Law*, *16*, 409–431.

Singer, P. W. and Cole, A. (2015). *Ghost fleet: A novel of the next World War*. New York: Houghton Mifflin Harcourt.

Smith, B. (1994). Fiat objects. In N. Guarino, S. Pribbenow and L. Vieu (eds), *Parts and wholes: Conceptual part–whole relations and formal mereology* (pp. 15–23). Proceedings of the European Conference on Artificial Intelligence. Amsterdam: ECAI. Retrieved from http://ontology.buffalo.edu/smith/articles/fiatobjects.pdf

Smith, S. A. (2014a). A Sino-Japanese clash in the East China Sea. *Contingency Planning Memorandum No. 18*. New York: Council on Foreign Relations (US). April. Retrieved from www.cfr.org/japan/sino-japanese-clash-east-china-sea/p30504

Smith, S. A. (2014b, 21 February). How China and Japan might avoid military confrontation in the East China Sea: *Asia Society*. Retrieved from http://asiasociety.org/blog/asia/how-china-and-japan-might-avoid-military-confrontation-east-china-sea

Smith, S. A. (2015). *Intimate rivals: Japanese domestic politics and a rising China*. New York: Columbia University Press.

Smits, G. (2015). Rethinking Ryukyu. *IJOS: International Journal of Okinawan Studies 6*, 1–19.

Soble, J. (2015, 18 September). Japan's Parliament approves overseas combat role for military. *New York Times*. Retrieved from www.nytimes.com/2015/09/19/world/asia/japan-parliament-passes-legislation-combat-role-for-military.html?emc=edit_th_20150919&nl=todaysheadlines&nlid=37589315&_r=0

Soble, J. (2016, 19 June). At Okinawa protest, thousands call for removal of U.S. bases. *New York Times*. Retrieved from www.nytimes.com/2016/06/20/world/asia/japan-okinawa-protest-united-states-military.html?emc=edit_th_20160620&nl=todaysheadlines&nlid=37589315&_r=0

Speetjens, J. (2002). *St. Martin yesterday, today*. Philipsburg, Sint Maarten: Foundation History St. Martin.

Spiliopoulou-Åkermark, S. (ed.) (2011). *The Åland example and its components: Relevance for international conflict resolution*. Mariehamn: The Åland Islands Peace Institute.

Spitsbergen Treaty (1920). *Treaty between Norway, The United States of America, Denmark, France, Italy, Japan, the Netherlands, Great Britain and Ireland and the British overseas Dominions and Sweden concerning Spitsbergen signed in Paris 9th February 1920*. Retrieved from http://emeritus.lovdata.no/traktater/texte/tre-19200209-001.html

State Council of the PRC (2012, September 25). *Diaoyu Dao, an inherent territory of China*. Beijing, China: State Council Information Office. Retrieved from http://news.xinhuanet.com/english/china/2012-09/25/c_131872152.htm

States of Jersey (2011). Tidal power for Jersey: the next steps. Recommendations to the Minister for Planning and Environment from the Tidal Power Commission. St Helier, Jersey: States of Jersey.

States of Jersey (2012a). *Les Écréhous and Les Dirouilles Ramsar Management Plan*. Trinity, Jersey: Department of the Environment.

States of Jersey (2012b). *Les Minquiers Ramsar Management Plan*. Trinity, Jersey: Department of the Environment.

Steijn, R. and Huizer, T. (2013), *OURCOAST experiences related to governance of the Wadden region*. Leeuwarden: Waddenacademie.

Stephan, J. J. (1974). *The Kuril islands: Russo-Japanese frontier in the Pacific*. Oxford: Oxford University Press.

Stevenson, C. (2007). Hans off! The struggle for Hans Island and the potential ramifications for international border dispute resolution. *Boston College International and Comparative Law Review*, *30*(1), 263–275.

Stewart, S. (1993). On *longing: Narratives of the miniature, the gigantic, the souvenir, the collection*. Durham, NC: Duke University Press.

Stuart, K. (2008). A global listing of sub-national island jurisdictions. In G. Baldacchino and K. Stuart (eds), *Pulling strings: Policy insights for Prince Edward Island from other sub-national island jurisdictions* (pp. 173–186). Charlottetown: Island Studies Press.

Stubbs, R. (2014). Strengthening the efficacy of acquisitive prescription in international law: implications for the Senkaku–Diaoyu islands dispute. *BYU Prelaw Review*, *28*(1), 49–64.

Suganuma, U. (2000). *Sovereign rights and territorial space in Sino-Japanese relations: Irredentism and the Diaoyu/Senkaku islands*. Honolulu, HI: University of Hawai'i Press.

Szonyi, M. (2008). *Cold War island: Quemoy on the front line*. Cambridge: Cambridge University Press.

Tadashi, I. (2013, 26 November). Getting Senkaku history right. *The Diplomat*. Retrieved from http://thediplomat.com/2013/11/getting-senkaku-history-right/

Taiwan–Japan Fisheries Agreement (2013). Embodying the ideals and spirit of the East China Sea Peace Initiative. Illustrated pamphlet. Retrieved from http://www.mofa.gov.tw/Upload/WebArchive/979/The%20Taiwan-Japan%20Fisheries%20Agreement%20(illustrated%20pamphlet).PDF

Takagi, T. (2016). Securing a stable supply of critical raw metals. *Synthesiology* (English edition), *9*(1), 16–26.

Takahashi, K. (2014, 13 February). Shinzo Abe's nationalist strategy. *The Diplomat*. Retrieved from http://thediplomat.com/2014/02/shinzo-abes-nationalist-strategy/

Takenaka, K. (2012, 11 September). Japan buys disputed islands, China sends patrol ships. Reuters. Retrieved from www.reuters.com/article/2012/09/11/us-japan-china-idUSBRE88A0GY20120911

Tanaka, S. (2015). The micro-foundations of territorial disputes: evidence from a survey experiment in Japan. *Conflict Management and Peace Science*. Retrieved from http://cmp.sagepub.com/content/early/2015/07/21/0738894215581330.full.pdf+html

Tansill, C. C. (1968). *The purchase of the Danish West Indies*. Westport, CT: Greenwood Publishing.

Taubenfeld, H. J. and Jessup, P. C. (1959). *Controls for Outer Space and the Antarctic analogy*. New York: Columbia University Press.

Taubman, P. (1984, 8 February). Shultz makes quick visit to Grenada. *New York Times*. Retrieved from www.nytimes.com/1984/02/08/world/shultz-makes-quick-visit-to-grenada.html

Taylor Fravel, M. (2008). *Strong borders, secure nation: Cooperation and conflict in China's territorial disputes*. Princeton, NJ: Princeton University Press.

Teller, M. (2014, 20 June). Why do so many nations want a piece of Antarctica? *BBC News*. Retrieved from www.bbc.com/news/magazine-27910375

Tenembaum, Y. J. (2012, 24 July). Is there a solution to the Falklands/Malvinas conflict? *MercoPress*. Retrieved from http://en.mercopress.com/2012/07/24/is-there-a-solution-to-the-falklands-malvinas-conflict

Tertre, J.-B. du (1667). *Histoire générale des Antilles habitées par les François*. Paris, France: Thomas Jolly (in French).

The Constitution of Finland (1999, 11 June). (731/1999). Retrieved from http://www.finlex.fi/fi/laki/kaannokset/1999/en/19990731.pdf

The Economist (2006, 25 May). Puerto Rico: trouble on Welfare Island. Retrieved from www.economist.com/node/6980051

The Economist (2008, 24 July). The cockerel's cropped crest: nearly 40 years after fighting flared, a border deal is reached. Retrieved from www.economist.com/node/11792951

The Economist (2012a, 22 September). Could Asia really go to war over these? Retrieved from www.economist.com/node/21563316

The Economist (2012b, 22 December). The Senkaku or Diaoyu Islands: narrative of an empty space. Retrieved from www.economist.com/news/christmas/21568696-behind-row-over-bunch-pacific-rocks-lies-sad-magical-history-okinawa-narrative

The Economist (2013, 31 August). Sea change: a gathering of Pacific leaders worries about climate change. Retrieved from www.economist.com/news/asia/21584396-gathering-pacific-leaders-worries-about-climate-change-sea-change

The Economist (2014a, 15 November). China and Japan: out of the deep freeze. Retrieved from www.economist.com/news/china/21632637-welcomeand-long-overduedetente-out-deep-freeze

The Economist (2014b, 6 December). The world is Xi's oyster. *Banyan*. Retrieved from www.economist.com/news/china/21635516-confident-chinese-leader-sets-out-his-foreign-policy-store-it-not-wholly-comforting-world

The Economist (2015a, 14 February). American forces in Japan: showdown. Retrieved from www.economist.com/news/asia/21643232-long-running-struggle-over-new-military-base-coming-head-showdown

The Economist (2015b, 28 February). Reclamation marks. Retrieved from www.economist.com/news/asia/21645268-unprecedented-building-boom-reclamation-marks

The Economist (2015c, 28 February). Such quantities of sand. Retrieved from www.economist.com/news/asia/21645221-asias-mania-reclaiming-land-sea-spawns-mounting-problems-such-quantities-sand

The Economist (2015d, 25 April). Japan and the United States: base issues. Retrieved from www.economist.com/news/asia/21649538-bilateral-alliance-has-served-well-decades-needs-makeover-base-issues

The Economist (2015e, 13 June). Revising Japan's constitution: peacemongers. Retrieved from www.economist.com/news/asia/21654122-new-defence-plans-run-difficulties-peacemongers

The Economist (2015f, 18 July). The South China sea: see u in court. Retrieved from www.economist.com/news/asia/21657807-tribunal-ponders-chinas-claims-see-u-court?fsrc=nlw|newe|20-07-2015|E

The Economist (2016a, 2 April). Chairman of everything. Retrieved from www.economist.com/news/china/21695923-his-exercise-power-home-xi-jinping-often-ruthless-there-are-limits-his

The Economist (2016b, 16 July). Taiwanese identity: Hello Kitty, goodbye panda. Retrieved from www.economist.com/news/asia/21702223-taiwans-obsession-japanese-kawaii-culture-hello-kitty-goodbye-panda

The Royal Forums (2013, 6 October). Oldest to newest monarchy in the world? Retrieved from www.theroyalforums.com/forums/f12/oldest-to-newest-monarchy-in-the-world-35253.html

Thomas, W. I. and Thomas, D. S. (1928). *The child in America: Behaviour problems and programs*. New York: Knopf.

Tiezzi, S. (2014, 19 April). Japan to station troops on Yonaguni, near disputed islands. *The Diplomat*. Retrieved from http://thediplomat.com/2014/04/japan-to-station-troops-on-yonaguni-near-disputed-islands/

Tiezzi, S. (2015, 8 January). Japan and China spar online over Senkaku/Diaoyu islands. *The Diplomat*. Retrieved from http://thediplomat.com/2015/01/japan-and-china-spar-online-over-senkakudiaoyu-islands/

Time (1968, 28 June). Churches: the concept of sanctuary. Retrieved from www.time.com/time/magazine/article/0,9171,841355,00.html

Ting, Shi (2016, 2 June). Two weeks into presidency, Taiwan leader pushes away from China. *Bloomberg News*. Retrieved from www.bloomberg.com/news/articles/2016-06-02/two-weeks-into-presidency-taiwan-leader-pushes-away-from-china

Today (2012, 16 February). St. Maarten and St. Martin celebrate 'everlasting cooperation'. Retrieved from www.todaysxm.com/2012/02/16/st-maarten-and-st-martin-celebrate-everlasting-cooperation/

Today (2014, 28 October). Committee hints at border controls and visa requirements: French want to dump Treaty of Concordia. Retrieved from http://www.todaysxm.com/2014/10/28/committee-hints-at-border-controls-and-visa-requirements-french-want-to-dump-treaty-of-concordia/

Toonen, Th. A. J. and Staatsen, A. A. M. F. (2004), *Goed bestuur voor de Wadden: Een essay over de organisatie van duurzaamheidsbesluitvorming*. Retrieved from www.waddenzee.nl/fileadmin/content/Dossiers/Overheid/pdf/Meijer_essay.pdf (in Dutch).

Trau, A. M. (2012). Beyond pro-poor tourism: (re)interpreting tourism-based approaches to poverty alleviation in Vanuatu. *Tourism Planning & Development, 9*(2), 149–164.

Treaty between Australia and Indonesia (1989, 11 December). *Treaty between Australia and the Republic of Indonesia, on the zone of cooperation in an area between the Indonesian Province of East Timor and Northern Australia*. Entered into force 9 February 1991. Retrieved from www.atns.net.au/agreement.asp?EntityID=710

Treaty between Australia and Timor Leste (2006, 12 January). *Treaty between Australia and the Democratic Republic of Timor-Leste on certain maritime arrangements in the Timor Sea*. Entered into force 23 February 2007. Retrieved from www.austlii.edu.au/au/other/dfat/treaties/2007/12.html

Treaty of Peace and Friendship Map (1984). Retrieved from https://en.wikipedia.org/wiki/Treaty_of_Peace_and_Friendship_of_1984_between_Chile_and_Argentina#/media/File:Beagle.1984.en.png

Tretiak, D. (1978). The Sino-Japanese treaty of 1978: the Senkaku incident prelude. *Asian Survey, 18*(12), 1235–1249.

Trimble, W. R. (1946). The embassy chapel question: 1625–1660. *Journal of Modern History, 18*(2), 97–107.

Tritten, T. J. and Sumida, C. (2012, 6 November). US sailors on Okinawa charged with gang rape, robbery. *Stars and Stripes*. Retrieved from www.stripes.com/news/pacific/okinawa/us-sailors-on-okinawa-charged-with-gang-rape-robbery-1.196116

Tyner, C. (2011, 9 March). A short history of feather hunting on Uotsurijima, Diaoyu/Senkaku islands. Retrieved from https://colintyner.wordpress.com/2011/03/09/a-short-history-of-feather-hunting-on-uotsurijima-diaoyusenkaku-islands/

Ugly Duck (2012). Hans across the water: make beer, not war. Ugly Duck Brewing Co. Retrieved from http://fr.myeurop.info/2012/07/24/faites-la-biere-pas-la-guerre-5728

UK–France Agreement (2000a). Agreement between the United Kingdom of Great Britain and Northern Ireland and the French Republic concerning the establishment of a maritime boundary between France and Jersey. St Helier, 4 July.

UK–France Agreement (2000b). Agreement between the Government of the United Kingdom of Great Britain and Northern Ireland and the Government of the French Republic concerning Fishing in the Bay of Granville, with Exchanges of Notes and Declaration. St Helier, 4 July.

UN (2006). Reports of International Arbitral Awards: Island of Palmas case (Netherlands, USA), 4 April 1928. *UN Cases, 2* (pp. 829-871). Retrieved from http://legal.un.org/riaa/cases/vol_II/829-871.pdf

UN Office for Outer Space Affairs (2015). *Treaty on principles governing the activities of states in the exploration and use of outer space, including the moon and other celestial bodies*. RES 2222 (XXI), 1966. Retrieved from http://www.unoosa.org/pdf/gares/ARES_21_2222E.pdf

UNCLOS (1982). *United Nations Conference on the Law of the Sea*. New York: UN Division for Ocean Affairs and Law of the Sea. Retrieved from http://www.un.org/depts/los/reference_files/chronological_lists_of_ratifications.htm#The United Nations Convention on the Law of the Sea

UNCLOS (1998). *The United Nations Convention on the Law of the Sea: A historical perspective.* New York: UN Division for Ocean Affairs and Law of the Sea. Retrieved from http://www.un.org/depts/los/convention_agreements/convention_historical_perspective.htm#Historical%20Perspective

UNEP (2015). *The global commons.* United Nations Environment Programme. Retrieved from www.unep.org/delc/GlobalCommons/tabid/54404/

UNFPA (2014). *Population and development profiles: Pacific island countries.* Suva: Pacific Sub-regional Office, United Nations Population Fund.

Unger, R. W. (1980). *The ship in the medieval economy: 600–1600.* Montreal and Kingston: McGill-Queen's University Press.

United Nations (2002). *Treaty of Peace and Friendship between Chile and Argentina* (with annexes and maps), 29 November 1984. Retrieved from www.un.org/Depts/los/LEGISLATIONANDTREATIES/PDFFILES/TREATIES/CHL-ARG1984PF.PDF

United Nations (2006a). *Treaty series*, Vol. 2269. New York: United Nations.

United Nations (2006b). Reports of International Arbitral Awards. Dispute between Argentina and Chile concerning the Beagle Channel. Volume XXI (pp. 53–264). Retrieved from http://legal.un.org/riaa/cases/vol_XXI/53-264.pdf

US Congress (2015, 15 June). *Space resource exploration and utilisation act of 2015*. H.R.1508-114th Congress (2015–2016). Retrieved from www.congress.gov/bill/114th-congress/house-bill/1508

US Supreme Court (1998). *New Jersey v. New York* 523 U.S. 767. Retrieved from https://supreme.justia.com/cases/federal/us/523/767/

USFWS (2008). Short-tailed albatross recovery plan. Anchorage AK: U.S. Fish and Wildlife Service. Retrieved from http://ecos.fws.gov/docs/recovery_plan/090520.pdf.

Valencia, M. J. (2007). The East China Sea dispute: context, claims, issues, and possible solutions. *Asian Perspective*, *31*(1), 127–167.

Van Aert, P. (2013). Tierra del Fuego. In G. Baldacchino (ed.), *The political economy of divided islands: Unified geographies, multiples polities* (pp. 195–212). New York: Palgrave Macmillan.

Van Der Merwe, C. (2015). *European condominium law*. Cambridge: Cambridge University Press.

Van Dijk, J., Broersma, L. and Mehnen, N. (2016). Options for socioeconomic developments in ICZM for the tri-national Wadden area. *Ocean & Coastal Management*, *119*(1), 76–92.

Van Dyke, J. M., Pettit, T. N., Clark, J. C. and Clark, A. L. (1988). Legal status of Johnston Atoll and its exclusive economic zone, *University of Hawaii Law Review*, *10*, 183–194.

Van Leeuwe, M. and Haartsen, T. (2012). Dynamics as key value of the Wadden Sea Region. *Ocean and Coastal Management*, *68*(1), 1–3.

Van Trease, H. (1987). *The politics of land in Vanuatu: From colony to independence*. Suva: Institute of Pacific Studies, University of the South Pacific.

Vaughan, A. (2015, 18 March). Pitcairn Islands to get world's largest single marine reserve. *Guardian*. Retrieved from www.theguardian.com/environment/2015/mar/18/pitcairn-islands-marine-reserve-budget-2015

Vézina, V. (2014). The role of the political system in shaping island nationalism: a case-study examination of Puerto Rico and Newfoundland. *Island Studies Journal*, *9*(1), 103–122.

Vine, D. (2011). *Island of shame: The secret history of the US military base on Diego Garcia*. Princeton, NJ: Princeton University Press.

Vladi Private Islands (2016). Île Gagnon. Retrieved from www.privateislandsonline.com/islands/ile-gagnon

Vlcek, W. (2008). Competitive or coercive? The experience of Caribbean offshore financial centres with global governance. *The Round Table: Commonwealth Journal of International Affairs*, *97*(396), 439–452.

Vollmer, M., Guldberg, M., Maluck, M., Marrewijk, D. and Schlicksbier, G. (2001). *Landscape and cultural heritage in the Wadden Sea region*. Project Report, Wadden Sea Ecosystem No. 12. Common Wadden Sea Secretariat, Trilateral Monitoring and Assessment Group, Wilhelmshaven.

Vuille-Kowing, K. (2011). Hans island: an Arctic border dispute between Canada and Denmark. *ICE Case Studies No. 272*, December. Retrieved from www1.american.edu/ted/ICE/kalaallit-nunaat.html

Wadden Sea Forum (2016). Wadden Sea Forum's region. Retrieved from www.waddensea-forum.org/en/forum/wsf-region

Waddenzeecommissie (1974). *Rapport van de Waddenzeecommissie*. The Hague: Rijkswaterstaat (in Dutch).

Walker, J. B. (2015). Archipelagic ambiguities: the demarcation of modern Japan, 1868–1879. *Island Studies Journal, 10*(2), 197–222.

Wang, C. (2014, 12 February). MAC Minister Wang in historic meeting. *Taipei Times*. Retrieved from www.taipeitimes.com/News/front/archives/2014/02/12/2003583307

Wang, Guanzhong (2014, 18 August). On Asian security concept. Deputy Chief of General Staff of the Chinese People's Liberation Army at the 13th Shangri-La Dialogue, 1 June. *English News (China)*. Retrieved from http://news.xinhuanet.com/mil/2014-09/18/c_127000792.htm

Wang, Vincent Wei-cheng (2015). The U.S. Asia rebalancing and the Taiwan Strait rapprochement, *Orbis, 59*(3), 361–379.

Wang, Z. (2014). *Never forget national humiliation: Historical memory in Chinese politics and foreign relations*. New York: Columbia University Press.

Want China Times (2015, 8 March). Taiwan, Japan agree on amended fishing regulations. Retrieved from www.wantchinatimes.com/news-subclass-cnt.aspx?id=20150308000110&cid=1101

Warnock, E. (2012, 1 September). Mayor of Ishigaki: disputed islands 'must be protected'. *Wall Street Journal*. Retrieved from http://blogs.wsj.com/japanrealtime/2012/09/01/mayor-of-ishigaki-disputed-islands-must-be-protected/

Watanabe, Y. and McConnell, D. L. (2015). *Soft power superpowers: Cultural and national assets of Japan and the United States*. New York: Routledge.

Watts, R. W. (2008). Island jurisdictions in comparative constitutional perspective. In G. Baldacchino and D. Milne (eds), *The case for non-sovereignty: Lessons from sub-national island jurisdictions* (pp. 25–44). London: Routledge.

Waugh, R. (2012, 11 July). Apple quietly gets rid of racist iPad game with Japanese invaders attacking disputed Chinese islands. *Mail On-Line*. Retrieved from www.dailymail.co.uk/sciencetech/article-2172095/Apple-quietly-gets-rid-controversial-iPad-game-Japanese-invaders-attacking-disputed-Chinese-islands.html

Weaver-Hightower, R. (2007). *Empire islands: Castaways, cannibals, and fantasies of conquest*. Minneapolis, MN: University of Minnesota Press.

Weinstein, M. (2006). South Korea–Japan Dokdo/Takeshima dispute: toward confrontation. *The Asia-Pacific Journal: Japan Focus*. Retrieved from www.japanfocus.org/-Michael-Weinstein/1685/article.html

Weisman, A. (2008). *The world without us*. New York: Palgrave Macmillan.

Weitz, R. (2008, 1 August). Russia, China end decades-long border dispute. *World Politics Review*. Retrieved from www.worldpoliticsreview.com/articles/2517/russia-china-end-decades-long-border-dispute

Weller, M. and Wolff, S. (eds) (2005). *Autonomy, self-governance and conflict resolution*. London: Routledge.

Welsh, T. (2015, 10 July). Dominican Republic temporarily halts deportation of Haitians. *US News*. Retrieved from www.usnews.com/news/articles/2015/07/10/dominican-republic-temporarily-halts-deportation-of-haitians

White, H. (2012). *The China choice: Why America should share power*. Melbourne, VIC: Black Inc Books.

Whittaker, R. J. and Fernández-Palacios, J. M. (2007). *Island biogeography: Ecology, evolution, and conservation*. Oxford: Oxford University Press.

Willis, D. B. (2009). Dejima: creolisation and enclaves of difference in transnational Japan. In D. B. Willis and S. Murphy-Shigematsu (eds), *Transcultural Japan: At the borderlands of race, gender and identity* (pp. 239–263). New York: Routledge.

Winterbottom, V. (2014, 10 February). China eyeing the future with fourth Antarctic research base. *The Interpreter*. Retrieved from www.lowyinterpreter.org/post/2014/02/10/China-expands-presence-in-Antarctica.aspx#.UvgRqEYbL5c.twitter

Wittmeyer, A. (2012, 24 September). The even smaller rocks Japan and China are fighting over. *Foreign Policy*. Retrieved from http://foreignpolicy.com/2012/09/24/the-even-smaller-rocks-japan-and-china-are-fighting-over/

Wolff, W. J. (1976). Introduction to the subject and goal of the conference of Wadden Sea experts. In W. J. Wolff (ed.), *Proceedings of the conference of Wadden Sea experts. Schiermonnikoog, The Netherlands, 26/28 November 1975* (pp. 29–33). The Hague: Staatsuitgeverij.

Wolff, W. J. (1992). The end of a tradition: 1000 years of reclamation of wetlands in The Netherlands come to an end. *Ambio*, *21*, 287–291.

Wolff, W. J. (2013). Ecology of the Wadden Sea: research in the past and challenges for the future. *Journal of Sea Research*, *82*, 3–9.

Wong, A. I. and Siu, R. S. (2015). Casino tourism: how does entrepreneurship change Macau? In G. Baldacchino (ed.), *Entrepreneurship in small island states and territories* (pp. 218–232). New York: Routledge.

Wong, E. and Buckley, C. (2015, 4 March). China's military budget increasing 10% for 2015, official says. *New York Times*. Retrieved from www.nytimes.com/2015/03/05/world/asia/chinas-military-budget-increasing-10-for-2015-official-says.html?_r=0

Wong, J. and Lye, L. F. (2014). Reviving the ancient Silk Road: China's new diplomatic initiative. *East Asian Policy*, *6*(3), 5–15.

Wong, J. Y. W. and Twitchett, D. (2002). *Deadly dreams: Opium and the Arrow war (1856–1860) in China*. Cambridge: Cambridge University Press.

Worsley, P. (1957). *The trumpet shall sound: A study of 'cargo' cults in Melanesia*. London: MacGibbon & Kee.

WSF (2010). *Without frontiers: Achievements in cross-border, cross-sector, communication and cooperation*. Wilhelmshaven: Wadden Sea Forum (WSF).

WSF (2013). *ICZM Strategy for the Wadden Sea region*. Wilhelmshaven: Wadden Sea Forum (WSF).

WTTC (2011). *The Tohoku Pacific earthquake and tsunami: Impact on travel & tourism*. World Travel and Tourism Council. Retrieved from www.wttc.org/-/media/files/reports/special%20and%20periodic%20reports/japan_report_may_2011_-_report_1_of_4.pdf

Xiaoming, Zhang (2015). China's perceptions of and responses to Abe's foreign policy. *Asian Perspective*, *39*(3), 423–440.

Xinhua (2013, 23 November). Announcement of the aircraft identification rules for the East China Sea Air Defense Identification Zone of the P.R.C. Retrieved from http://news.xinhuanet.com/english/china/2013-11/23/c_132911634.htm

Yamaguchi, E. and Arakawa, Y. (2002). *The demise of the Ryukyu kingdom: Western accounts and controversy*. Okinawa: Yojushorin.

Yangwen, Z. (2005). *The social life of opium in China*. Cambridge: Cambridge University Press.

Yi Fu Tuan (2014, 4 April). *Space, place and nature: The farewell lecture*. Madison, WI: University of Wisconsin-Madison. Retrieved from www.yifutuan.org/dear_colleague.htm

Yokell, M. A. (2010). *The treaty of Helgoland–Zanzibar: The beginning of the end for the Anglo-German friendship?* Doctoral dissertation. Richmond, VA: University of Richmond.

Yonetani, J. (2000). Ambiguous traces and the politics of sameness: placing Okinawa in Meiji Japan. *Japanese Studies*, *20*(1), 15–31.

Yoshida, T. (2006). *The making of the 'Rape of Nanking': History and memory in Japan, China, and the United States*. Oxford: Oxford University Press.

Zachs, R. (2005). *The pirate coast: Thomas Jefferson, the first Marines, and the secret mission of 1805*. London: Hachette.

Zang, D. M. (1991). Frozen in time: the Antarctic Mineral Resource Convention. *Cornell Law Review*, *76*(3), Article 4. Retrieved from http://scholarship.law.cornell.edu/cgi/viewcontent.cgi?article=3496&context=clr

Zazzle (2015). Hans island T-shirt. Retrieved from www.zazzle.ca/hans_o_hans_island_the_nationalistic_version_tshirt-235190314210203467

Zhang, J. J. (2010). Of Kaoliang, bullets and knives: local entrepreneurs and the battlefield tourism enterprise in Kinmen (Quemoy), Taiwan. *Tourism Geographies*, *12*(3), 395–411.

Zhang, Yan (2015, 17 September). UN peacekeepers to help police fight crime. *China Daily*. Retrieved from www.chinadaily.com.cn/china/2015-09/17/content_21898965.htm

Zhao, Kejin and Gao, Xin (2015). Pursuing the Chinese dream: institutional changes of Chinese diplomacy under President Xi Jinping. *China Quarterly of International Strategic Studies*, *1*(1), 35–57. Retrieved from www.worldscientific.com/doi/abs/10.1142/S2377740015500037

Zheng, Xin (2012, 11 July). Diaoyu Islands game pulled from app site. *China Daily*. Retrieved from www.chinadaily.com.cn/china/2012-07/11/content_15566881.htm

Zhu, Feng (2014, 25 November). China's first ADIZ decision: One year later. *Asia Maritime Transparency Initiative*. Retrieved from http://amti.csis.org/chinas-first-adiz-decision-one-year-later/

Zilio, M. (2012, 26 November). Fate of disputed Arctic island up for debate this week in Ottawa: report. *iPolitics*. Retrieved from http://ipolitics.ca/2012/11/26/fate-of-disputed-arctic-island-up-for-debate-this-week-in-ottawa-report/

Zou, Keyuan (2006). Joint development in the South China Sea: a new approach. *The International Journal of Marine and Coastal Law*, *21*(1), 83–109.

ZQGame (2016). Retrieved from https://en.zqgame.com/

Discography

Anderson, B. and Ulvaeus, B. (1980). 'The winner takes it all'. First single from *Super Trouper* album, released by ABBA, 21 July. *Polar Music*.

Filmography

Kingdom of Heaven (2005). Ridley Scott (Producer and Director). Los Angeles: Twentieth Century Fox Productions.

Index

(Surnames in UPPER CASE indicate the names of the expert authors who contribute to this volume. Their page references in the index include the page ranges of their contributions.)

For Product Safety Concerns and Information please contact our EU representative GPSR@taylorandfrancis.com
Taylor & Francis Verlag GmbH, Kaufingerstraße 24, 80331 München, Germany

www.ingramcontent.com/pod-product-compliance
Ingram Content Group UK Ltd.
Pitfield, Milton Keynes, MK11 3LW, UK
UKHW011001180926
475701UK00004B/503
9780367218775